T0345282

# What Every Engineer Should Know About Cyber Security and Digital Forensics

Most organizations place a high priority on keeping data secure, but not every organization invests in training its engineers or employees in understanding the security risks involved when using or developing technology. Designed for the non-security professional, *What Every Engineer Should Know About Cyber Security and Digital Forensics* is an overview of the field of cyber security.

The Second Edition updates content to address the most recent cyber security concerns and introduces new topics such as business changes and outsourcing. It includes new cyber security risks such as Internet of Things and Distributed Networks (i.e., blockchain) and adds new sections on strategy based on the OODA (observe-orient-decide-act) loop in the cycle. It also includes an entire chapter on tools used by the professionals in the field. Exploring the cyber security topics that every engineer should understand, the book discusses network and personal data security, cloud and mobile computing, preparing for an incident and incident response, evidence handling, internet usage, law and compliance, and security forensic certifications. Application of the concepts is demonstrated through short case studies of real-world incidents chronologically delineating related events. The book also discusses certifications and reference manuals in the areas of cyber security and digital forensics.

By mastering the principles in this volume, engineering professionals will not only better understand how to mitigate the risk of security incidents and keep their data secure, but also understand how to break into this expanding profession.

# What Every Engineer Should Know

*Series Editor*
*Phillip A. Laplante*
*Pennsylvania State University*

What Every Engineer Should Know about Excel, Second Edition
*J.P. Holman and Blake K. Holman*

Technical Writing: A Practical Guide for Engineers, Scientists, and Nontechnical Professionals, Second Edition
*Phillip A. Laplante*

What Every Engineer Should Know About the Internet of Things
*Joanna F. DeFranco and Mohamad Kassab*

What Every Engineer Should Know about Software Engineering
*Phillip A. Laplante and Mohamad Kassab*

What Every Engineer Should Know About Cyber Security and Digital Forensics
*Joanna F. DeFranco and Bob Maley*

For more information about this series, please visit: www.routledge.com/What-Every-Engineer-Should-Know/book-series/CRCWEESK

# What Every Engineer Should Know About Cyber Security and Digital Forensics

## Second Edition

Joanna F. DeFranco and Bob Maley

CRC Press
Taylor & Francis Group
Boca Raton  London  New York

CRC Press is an imprint of the
Taylor & Francis Group, an **informa** business

Second edition published 2023
by CRC Press
6000 Broken Sound Parkway NW, Suite 300, Boca Raton, FL 33487-2742

and by CRC Press
4 Park Square, Milton Park, Abingdon, Oxon, OX14 4RN

*CRC Press is an imprint of Taylor & Francis Group, LLC*

© 2023 Joanna F. DeFranco and Bob Maley

First edition published by CRC Press 2014

ISBN: 978-1-032-14601-0 (hbk)
ISBN: 978-1-032-15665-1 (pbk)
ISBN: 978-1-003-24522-3 (ebk)

DOI: 10.1201/9781003245223

Typeset in Times
by codeMantra

# Contents

# What Every Engineer Should Know: Series Statement

What every engineer should know amounts to a bewildering array of knowledge. Regardless of the areas of expertise, engineering intersects with all the fields that constitute modern enterprises. The engineer discovers soon after graduation that the range of subjects covered in the engineering curriculum omits many of the most important problems encountered in the line of daily practice—problems concerning new technology, business, law, and related technical fields.

With this series of concise, easy-to-understand volumes, every engineer now has within reach a compact set of primers on important subjects such as patents, contracts, software, business communication, management science, and risk analysis, as well as more specific topics such as embedded systems design. These are books that require only a lay knowledge to understand properly, and no engineer can afford to remain uniformed of the fields involved.

# Preface

Long gone are the days where the security of your critical data could be protected by security guards, cipher locks, and an ID badge worn by all employees. As the computing paradigm is continually changing with shared resources, new technology and mobility, firewalls and anti-virus software are also not enough to protect critical assets.

This book will cover topics that range from the processes and practices that facilitate the protection from attacks, destruction, and unauthorized access of our private information and critical assets to the processes and practices that enable an effective response if and when the attacks, destruction, and unauthorized access occur. This book will provide information on those topics via real situations, case law, and the latest processes and standards from the most reliable sources. The goal is not for you to become a fully trained security or digital forensic expert (although we will explain how to accomplish that), but to provide accurate and sufficient information to pique your interest and to springboard you onto the right path if this is an area you wish to understand and/or pursue. If you're not aiming to be the next security professional at your company, this book can assist you in understanding the importance of security in your organization whether you are designing software, have access to personal data, or manage the day-to-day activities in your office, because we all need to take a part in protecting those critical assets. In any case, we are hoping this book will give you a new appreciation for the world of cyber security and digital forensics.

There are three main goals of this book. The first goal is to introduce the cyber security topics every engineer should understand. It is important to understand these topics, as most engineers work for organizations that need their data secure, and unfortunately not every organization invests in training their employees to understand how to reduce the risk of security incidents. It is a well-known fact that the weakest link in any system is the user. Just ask any hacker. The second goal is demonstrating the application of the security concepts presented. This will be accomplished by presenting case studies of real-world incidents. The final goal is to provide information on certifications in the area of cyber security and digital forensics for the reader who wants to take advantage of the vast and growing opportunities in this field.

# Acknowledgments

Many people provided invaluable support and assistance in various ways during the writing of this manuscript. I want to take this opportunity to thank the following people:

- Patrick Siewert, Founder & Principal Consultant of Pro Digital Forensic Consulting (Chapter 7).
- Patrick J. Burke, Counsel, Norton Rose Fulbright US LLP, and Susan Ross, Senior Counsel, Norton Rose Fulbright US LLP (Chapter 8).
- Dr. Ferhat Dikbiyik (contributions to Chapter 3).
- Special Agent Kathleen Kaderabek for her input regarding FBI training and the InfraGard organization, as well as for her comments on Chapter 8.
- Keith J. Jones, Senior Partner at Jones Dykstra & Associates, for sharing his experience on the high-profile case *U.S. v. Duronio*.
- Gabrielle Vernachio and Allison Shatkin at Taylor & Francis, for all of their assistance and encouragement throughout this project.
- Dr. Phillip Laplante, for his invaluable mentoring.

# Authors

**Joanna F. DeFranco,** earned her Ph.D. in computer and information science from New Jersey Institute of Technology, M.S. in computer engineering from Villanova University, and a B.S. in electrical engineering and math from Penn State University. She is an Associate Professor of Software Engineering at the Pennsylvania State University. She has worked as an Electronics Engineer for the Navy as well as a Software Engineer at Motorola. Dr. DeFranco is also a researcher for the National Institute of Standards and Technology (NIST) working with the Secure Systems and Applications group. She is a senior member of the IEEE and an area and column editor for *IEEE Computer Magazine*. Her research interests include software engineering, software security, distributed networks, and Internet of Things.

**Bob Maley,** Inventor, CISO, Author, Futurist, and OODA Loop fanatic, is the Chief Security Officer at Black Kite, the leader in third-party cyber risk intelligence.

Bob has been a leader in security for decades, initially in physical security as a law enforcement officer. He has acquired a broad range of experience and expertise in all areas of security, including third-party security, risk assessment, architecture, design, policy development, deployment, incident response and investigation, and enterprise solution deployments such as intrusion detection, data protection, compliance, and incident reporting and response.

Before joining Black Kite, Bob was the head of PayPal's Global Third-Party Security & Inspections team, developing the system into a state-of-the-art risk management program.

In a previous role as chief information security officer for the Commonwealth of Pennsylvania, he led the Pennsylvania Information Security Architecture program to win the 2007 award for outstanding achievement in information technology by the National Association of State Chief Information Officers (NASCIO).

Bob has been named a CSO of the Year finalist for the SC Magazine Awards and was nominated as the Information Security Executive of the Year, North America. Additionally, his team was a finalist in the SC Magazine Awards for Best Security Team. Bob's certifications include CRISC, CTPRP, OpenFAIR, and CCSK. His expertise has been quoted in numerous articles for *Forbes, Politico, Payments.com, StateTech Magazine, SC Magazine, Wall Street Journal, Washington Post, Dark Reading*, etc.

## ERRORS

Despite our best effort as well as the effort of the reviewers and the publisher, there may be errors in this book. If errors are found, please report them to jfd104@psu.edu.

# 1 Security Threats

Human error is not an explanation, rather it is something to be explained. In analyzing and learning from incidents, not just security incidents, you should never be satisfied with anyone closing out a post-mortem or issue as having a root cause of human error.

*—Phil Venables, Google Cloud CISO*

## 1.1 INTRODUCTION

If you use a computer that is connected to the Internet, your information is at risk. Our sensitive data can also be stolen through third-party vendors. Even more startling is that the time between an attack and disclosure from a third party is about 75 days! Imagine what can be done with stolen information in that amount of time. In addition, the most common type of third-party attack in 2021 was found to be ransomware (i.e., malicious software designed to block access to a computer until a sum of money is paid). In fact, these ransomware attacks initiated 27% of the breaches analyzed in 2021 (Black Kite 2022). In Black Kite's *2022 Third Party Breach Report*, they also found that the healthcare industry was the most common victim of attacks—accounting for 33% of the incidents in 2021, and software publishers were the most common source of third-party breaches.

Even if you are not an engineer working at a business that is considered critical infrastructure or a company that has a more moderate risk level, you have an identity and personal information that you need to protect; thus, you need to be an informed computer user.

The Internet Crime Complaint Center (IC3), a partnership between the Federal Bureau of Investigation (FBI) and the National White Collar Crime Center (NW3C), reports close to 800K Internet crime complaints in 2020—up from 300K in 2019! As a reference point in 2011 there were only about 300K complaints a year from the same report! (Internet Crime Report 2020). A few of the crimes reported include phishing scams, non-payment/non-delivery scams, and extortion. These numbers indicate that you need to prepare yourself and your business for an attack—because it will happen eventually.

Why are these attacks so much more prevalent and sophisticated? Because, as shown in Figure 1.1, the technical knowledge required by the hacker is decreasing. The attacks listed only highlight a few types of vulnerabilities, but there are enough shown to verify the point that it does not take a PhD or 20 years of computer experience to hack into a computer. The FBI has knocked on the doors of many people who are the parents of the "model" teenager. In a particular case, the teenager who was known for just hanging out at home and using the family computer but was actually hacking into NASA's computers.[1]

DOI: 10.1201/9781003245223-1

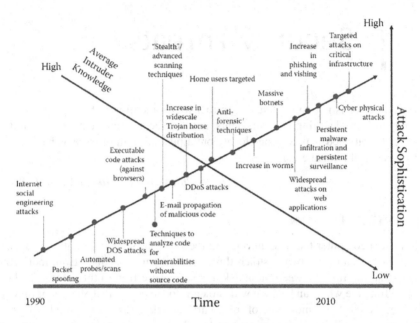

**FIGURE 1.1**   The trends in cyber-attacks. (Adapted from Lipson, H., 2002, special report CMU/SEI-2002-SR-009, and Carnegie Mellon, 2010, http://www.cert.org/tces/pdf/archie%20 andrews.pdf.)

Consider the following:

- The ScanSafe Annual Global Threat Report recorded a 252% growth in attacks on banking and financial institutions, 322% growth in attacks on pharmaceutical and chemical industries, and 356% growth in attacks on the critical oil and energy sectors in 2009 (www.scansafe.com/downloads/gtr/2009_AGTR.pdf).
- More than half of the operators of power plants and other critical infrastructure suspect that foreign governments have attacked their computer networks (Baker 2010).
- Of those operators, 54% acknowledged they had been hit by stealthy infiltration—applications planted to steal files, spy on e-mails, and control equipment inside a utility (Baker 2010).
- At nearly 2,500 companies, such as Cardinal Health and Merck, 75,000 computer systems have been hacked by malicious "bots" that enabled the attacker to manipulate the user's computer and steal personal information (Nakashima 2010).

New threats are constantly being reported, largely on the infrastructure of only a few countries. The attacks on these systems often exploit vulnerabilities provided by unwary users—and we can all be "unwary users" at times.

The focus and goal of this chapter are to highlight some of the common cyber security risks. We will start with the one that is the most difficult to defend against: social engineering. It is difficult to defend against because it preys on human nature to want to be helpful and kind. Once the social engineer finds a victim, he or she just needs to persuade (trick) the victim into revealing information that will compromise the security of the system.

Cause for Paranoia?[2]

## 1.2 SOCIAL ENGINEERING

The greatest threat to the security of your business is the social engineer (Mitnick and Simon 2002). In other words, your company can employ the latest state-of-the-art security equipment and it will still be vulnerable due to the ignorance of the system's users. Essentially, the social engineer takes advantage of the weakest link in your company—the user (see Figure 1.2). They are able to obtain confidential information without the use of technology.

The confidential information obtained by the social engineer is used to perform fraudulent activities or gain unauthorized access to a computer system. As you

FIGURE 1.2 The weakest link in the company. (Weiner, Z., 2012, Hacking, http://www.smbc-comics.com/, February 20, 2012.)

can imagine, social networking has made social engineering even easier. In an interview with Kevin Mitnick, the person who made social engineering famous, he described using a "spear phishing" tactic where an e-mail targets a specific person or organization coming from a trusted source. The person is targeted using information found on a social networking site. For example, the social engineer goes to *LinkedIn* and looks for network engineers because they usually have admin rights to the network (Luscombe 2011). Then, he or she sends those network engineers an e-mail (since he or she knows where they work) or calls them to obtain the needed information. Even a company specializing in cyber-attack recovery is a spear phishing target. In a report written by Mandiant (2013), a spear phishing attack was described targeting the company's CEO, Kevin Mandia. The goal was to attack the organization with an advanced persistent threat (APT[3]). The spear phishing e-mail was sent to all Mandiant employees. The e-mail was spoofed to appear as if it came from the company's CEO, Mr. Mandia. The e-mail, shown in Figure 1.3, had a malicious APT attachment (notice the spoofed e-mail address: @ rocketmail.com).

A criminal can always obtain user credentials purchased from the dark web (i.e., Internet content requiring specific software and configurations to access) and perform a "credential stuffing" attack. Credential stuffing is the automatic injection of stolen username and password pairs into website login forms to gain unauthorized account access (owasp.org). For example, the hack that compromised the largest fuel pipeline in the US and caused fuel shortages across the East Coast was a result of a single compromised password—suspected to have been retrieved from a dark web leak (Turton and Mehrotra 2021). However, the easiest way to gain fraudulent account access is still using social engineering. To demonstrate how easy a social engineering attack is to execute, let us compare the steps a high-tech hacker and a no-tech hacker (social engineer) would use to get a password (Long 2008). As you read through the steps, keep in mind that it is estimated that the high-tech way takes about a week and the no-tech way takes merely a moment or two.

Date: Wed, 18 Apr 2012 06:31:41 -0700
From: Kevin Mandia kevin.mandia@rocketmail.com
Subject: Internal Discussion on the Press Release

Hello,
Shall we schedule a time to meet next week?
We need to finalize the press release.
Details click here.

Kevin Mandia

FIGURE 1.3   Spoofed e-mail. (Adapted from Mandiant APT1 report, 2013, www.mandiant. com.)

A summary of the five-step high-tech way to obtain a password:

1. *Strategically scan the company network:* In a stealthy manner (from several IP addresses), search for ports listening to the Internet.
2. *Install malware on a victim's machine:* Sneak the rootkit (malware) onto the open port.
3. *Enumerate the target network:* While continuing to hide your activity, determine the network topology; for example, the size of the network, number of switches, and the location of the servers.
4. *Locate and copy the encrypted password file:* Covertly take a copy of the network hashes to analyze on your own network. This may result in acquiring passwords.
5. *Run automated cracking tools against the encrypted password file:* Use the password hashes from step 4 with your favorite password cracking tool. Passwords (e.g., leaked credentials), originally stolen from a successful breach, can also show up and circulate on the dark web (https://blog.castle.io/leaked-credentials-database/).

A summary of the two-step no-tech way to obtain a password:

1. *Make a phone call:* Ask easy questions. Find a way to swindle the person who answered the phone to reveal information such as terminology that only the insiders utilize. You may even be able to convince the person to provide you with access—which would eliminate step 2 of this process!
2. *Make another phone call:* In this conversation, use the information from the first phone call. You will now seem like one of them and the person on the other end will want to help you login! Essentially, one piece of information helps you get more information.

What needs to be understood at this point is that sensitive information can be obtained by just asking for it. In essence, social engineers take advantage of our human nature of kindness, which makes it easy for the social engineer to pretend to be someone

### *NEWS OF THE WORLD* MOBILE PHONE HACKING SCANDAL

*News of the World*, a British tabloid, was put out of business after 168 years due to the ramifications of phone hacking allegations. The newspaper was accused of hacking the mobile phone voicemail of celebrities, politicians, members of the British Royal Family, and Milly Dowler, a murder victim. Hacking into Dowler's phone was considered evidence tampering, and the hackers could face about 500 civil claims (Sonne 2012). Most of the victims were hacked because the default PINs for remote voicemail access were never changed. Even if the user did change the PIN, the "hacker" used social engineering techniques to trick the operator into resetting the PIN (Rogers 2012).

else. Thus, when he or she is armed with a few pieces of information, more information to break into secure networks can easily be acquired.

In his book, *The Art of Deception,* Kevin Mitnick goes through story after story based on what he calls one of the fundamental tactics of social engineering: "gaining access to information that a company employee treats as innocuous, when it isn't" (Mitnick and Simon 2002). Social engineering tactics can only be countered by properly training the system users. Things have only gotten worse since then. In 2018, Mitnick said, "The ever-changing security landscape doesn't hinder social engineering hacks. It can even enable them, allowing for more complex and effective methods to gain information from secure systems by using the humans that run them" (https://www.mitnicksecurity.com/in-the-news/social-engineering-from-the-trojan-horse-to-firewalls).

## 1.3 TRAVEL

Do you or your colleagues travel abroad? Social engineering can also occur when traveling. Businesspeople, US government employees, and contractors that are traveling abroad are routinely targeted for a variety of sensitive information, shown in Table 1.1.

The targeting takes many forms, according to the "Report to Congress on Foreign Economic Collection and Industrial Espionage":

- Exploitation of electronic media and devices
- Secretly entering hotel rooms to search
- Aggressive surveillance
- Attempts to set up romantic entanglements

The exploitation could simply occur through software updates while using a hotel Internet connection (FBI E-scams 2012). A pop-up window will appear to update software while the user is establishing an Internet connection in the hotel room.

---

**TABLE 1.1**

**Sensitive Information Targeted by Foreign Collectors**

### Critical Business Information May Include

| | |
|---|---|
| Customer data | Phone directories |
| Employee data | Computer access protocols |
| Vendor information | Computer network design |
| Pricing strategies | Acquisition strategies |
| Technical components and plans | Investment data |
| Corporate strategies | Negotiation strategies |
| Corporate financial data | Passwords (computer, phone, accounts) |

*Source:* US Department of Justice, Federal Bureau of Investigation, n.d., business travel brochure.

---

If the pop-up is clicked, the malicious software is installed on the laptop. The FBI recommends either performing the upgrade prior to traveling or going directly to the software vendor's website to download the upgrade. All of these threats can be mitigated by training, as will be discussed in Chapter 4.

## 1.4   MOBILE DEVICES

Many people use mobile devices to conduct business. As smartphones have become more prevalent, the hackers have taken notice. McAfee reports an increase in mobile threats from approximately 2,000 in 2011 to more than 8,000 in 2012. In 2021, McAfee reports mobile malware in the millions. Part of the reason for the increase lies in McAfee's ability to detect these threats, but nonetheless, that is a significant amount of malware. At this point, most of the malware, usually contained in phone apps, targets the Android operating system because of the open-source environment. The Android OS has been targeted because it does not provide adequate control over the private data, which are misused by third-party smartphone apps (Enck et al. 2010). Researchers at Penn State, Duke, and Intel Labs (2010) created an app called TaintDroid to monitor the behavior of third-party smartphone applications. They found that out of 30 popular Android apps, there were 68 instances of private information misuse across 20 of the apps. For example, an innocent wallpaper app of a favorite character will send your personal information to China (Mokey 2010). There is a lot of pressure on developers to produce more functionality faster and at lower cost, which limits the time needed to improve mobile security (Hulburt, Voas, and Miller 2011). This is not to discourage smartphone use or app development, but rather to encourage awareness of the risks when downloading apps to your smartphone.

Is your iPhone a spiPhone? Researchers at Georgia Tech discovered how to use the smartphone accelerometer to sense computer keyboard vibrations and can decipher typing with 80% accuracy. The accelerometer is the internal device that detects phone tilting (Georgia Tech 2011). A possible attack scenario could be the user downloading a seemingly harmless application that includes the keyboard-detection malware. So, do not set your phone too close to your keyboard! Placing your phone 3 or more inches away from your keyboard is recommended.

## 1.5   INTERNET

The Internet is both a benefit and a detriment: It created a global transformation of our economy, but also threatens our privacy. According to McAfee Labs (2012), the amount of known malware applications is over 80 million and continues to grow. The usual problems are, of course, fake antivirus (alerting victims of threats that do not exist), AutoRun (exploits mostly via USB), and password stealing (malware monitoring keystrokes). But, of greatest concern are rootkits which provide stealthy remote access to live resources and remain active for long periods on your system.

The Federal Communication Commission's (FCC) chairman, Julius Genachowski, has stated that the three top cyber threats are botnets, domain name fraud, and Internet protocol route hijacking (Grace 2012). Bot-infected computers are computers that are controlled by an attacker. A botnet is the collection of those computers that, according to the FCC, "pose a threat to the vitality and resiliency of the Internet and the online economy." Domain name fraud converts the domain name (e.g., www.google.com) to an incorrect IP address, thus sending the user to a website where fraudulent activity will probably occur. Internet protocol hijacking is where the Internet traffic is redirected through untrustworthy networks. Mitigation tactics to these threats will be discussed later in this book.

## 1.6   THE CLOUD

The cloud model shares resources such as networks, servers, storage, applications, and services. In other words, a cloud offers computing, storage, and software "as a service" (Buyya, Broberg, and Goscinski 2010). According to the National Institute of Standards and Technology (NIST), a federal agency that provides standards to promote US innovation and industrial competitiveness, there are four varieties of clouds (Mell and Grance 2011):

1. A *private cloud*, where a single organization shares the resource infrastructure exclusively
2. A *community cloud*, where the users of the cloud infrastructure are from different organizations that share the same concerns (e.g., all of the organizations may need to consider the same security regulations)
3. A *public cloud*, where almost anyone can utilize its resources
4. A *hybrid cloud*, where the preceding three varieties are combined and connected to enable data and application sharing

No matter which variety of cloud you utilize, clouds essentially provide three types of services: infrastructure as a service (IaaS), platform as a service (PaaS), and software as a service (SaaS); see Table 1.2.

**TABLE 1.2**
**Cloud Services**

| Service Class | Service Content |
|---|---|
| SaaS | Cloud applications |
| | *Examples: social networks, office applications, video processing* |
| PaaS | Cloud platform |
| | *Examples: programming, languages, frameworks* |
| IaaS | Cloud infrastructure |
| | *Examples: data storage, firewall, computation services* |

*Source:*  Buyya, R. et al., 2010, *Cloud Computing Principles and Paradigms.* New York: John Wiley & Sons.

In addition to these cloud services, there are cloud services, related to security and privacy such as monitoring and addressing malware, spam, and phishing problems that come through e-mail. The cloud model is great, especially for small businesses that would not be able to provide an expensive, effective infrastructure without spending a lot of money. However, instead of money, the hefty price tag is the risk that comes with sharing these types of resources.

The "Guidelines on Security and Privacy in Public Cloud Computing," published by NIST (Jansen and Grance 2011), discusses four fundamental concerns of the cloud. First is system complexity. This complexity brings with it a large playground for attackers. The cloud offers so many services and sometimes even nest and layer services from other cloud providers. Combining this complexity with the necessity of upgrades and improvement, unexpected interactions are created along with opportunities for hackers. The second concern is the fact that components and resources are shared unknowingly with other consumers. Your data are separated "logically," not "physically." This shared multitenant environment creates another opportunity for someone to gain unauthorized access. A good example is a security breach that occurred with Google Docs that allowed users to see files that were not "owned" or "shared" by them (Kaplan 2008).

The third concern is the fact that applications that were utilized from the company Intranet are now used over the Internet, thus increasing network threats. And finally, by utilizing a cloud model, you have given the control of your information to the people who manage the cloud. This loss of control "diminishes the organization's ability to maintain situational awareness, weigh alternatives, set priorities, and affect changes in security and privacy that are in the best interest of the organization" (Jansen and Grance 2011).

The IBM Trend and Risk Report for 2011 also recognized the vulnerability that cloud computing brings to your systems. They suggest that when thinking about the risk of using a cloud infrastructure, you should consider the following questions:

- Has your security team audited the practices of your partners?
- Are the practices consistent with yours?
- How confident are you in their execution?

Even with these risks, companies still consider a cloud infrastructure because there are a few advantages (Jansen and Grance 2011):

1. The cloud providers are able to have staff that is highly trained in security and privacy.
2. The platform is more uniform, thereby enabling better automation of security management activities such as configuration control, vulnerability testing, security audits, and patching.
3. With the amount of resources available, redundancy and disaster recovery capabilities are built in. They are also able to handle increased demands as well as contain and recover more quickly from cyber-attacks.

4. Data backup and recovery can be easier because of the superior policies and procedures of the cloud provider.
5. The client of the cloud can be easily supported on mobile devices (laptops, notebooks, netbooks, smartphones, and tablets) because all of the heavy-duty computational resources are in the cloud.
6. The risk of theft and data loss is lowered (though not gone) because the data are maintained and processed at the cloud.

## 1.7    CYBER PHYSICAL SYSTEMS

Cyber physical systems (CPS) integrate cyber, computational, and physical components to provide mission-critical systems. Examples of systems are smart electricity grids, smart transportation, and smart medical technology. These systems are "smart" because they are able to collect and use sensitive information from their environment to have an effect on the environment. But, because of the vast applications of CPS and the integrated computers and networks, they are impacted by cyber security.

The CPS needs not only to be usable but also to be safe and secure because the loss of security for a CPS can "have significant negative impact including loss of privacy, potential physical harm, discrimination, and abuse" (Banerjee et al. 2011). The first step in securing a CPS system is being aware of the cyber-attacks that may impact the system. The smart grid, for example, needs to include the following security properties (Govindarasu, Hahn, and Sauer 2012):

1. *Confidentiality* and protection of the information from unauthorized disclosure
2. *Availability* of the system/information where it remains operational when needed
3. *Integrity* of the system/information from unauthorized modification
4. *Authentication* prior to access by limiting access only to authorized individuals
5. *Nonrepudiation,* where the user or system is unable to deny responsibility for a previous action

CPS is relatively young; thus, as these systems are being designed, we need to keep in mind the necessary components to uphold the security properties.

## 1.8    THEFT

You not only need to improve your security posture to protect against hackers, but you also need to monitor the activities of your own employees. It is difficult to imagine that someone you trusted enough to hire would steal from you, but as we know this happens every day. Consider a situation where making and selling a specific food product contributes to most of a company's revenue. None of the company's competitors have been able to duplicate this product. Thus, the recipe is guarded and only a

In the Hewlett-Packard 2012 "Cyber Risk Report," researchers determined the risk trends for cyber security. For example, the number of new disclosed vulnerabilities had increased by 19% from 2011. These come from every angle, such as web applications, legacy technology, and mobile devices. For example, the skyrocketing mobile device sales in 2012 brought with it a similar number of mobile application vulnerabilities. Mobile device applications alone have seen a 787% increase in vulnerability disclosures. Understanding a company's technical security risk begins with knowing how and where the vulnerabilities occur within the organization (Hewlett-Packard 2013).

few people have access to it. One of the people that know this trade secret announces that she is leaving but gives the impression that she is retiring. However, her plan is to work for a competitor. The security team determined from analyzing her system activity that she had begun accessing confidential files and storing them on a flash drive in the weeks prior to her departure.

Another example was described in the "Report to Congress on Foreign Economic Collection and Industrial Espionage." In this situation, an employee downloaded a proprietary paint formula valued at $20 million that he planned to deliver to his new employer in China. Just recently it was discovered at the University of South Carolina Health and Human Services that an employee e-mailed himself over 200,000 patient records. These examples show that sometimes it is the authorized users who cause the data breaches. There are many ways to protect against theft, which will be discussed in Chapters 3 and 4.

## REFERENCES

Baker, S. January 2010. In the crossfire: Critical infrastructure in the age of cyber war. McAfee. http://resources.mcafee.com/content/NACIPReport.

Banerjee, A., Venkatasubramanian, K., Mukherjee, T. and Gupta, S. 2012. Ensuring safety, security, and sustainability of mission-critical cyber-physical systems. *Proceedings of the IEEE* 100(1).

Black Kite. 2022. 2022 third-party breach report: Trends, root causes and lessons learned from 2021. https://blackkite.com/whitepaper/2022-third-party-breach-report/.

Buyya, R., Broberg, J. and Goscinski, A. 2010. *Cloud computing principles and paradigms.* New York: John Wiley & Sons.

Carnegie Mellon, Software Engineering Institute. November 2010. Trusted computer in embedded systems. http://www.cert.org/tces/pdf/archie%20andrews.pdf (retrieved May 1, 2012).

Enck, W., Gilbert, P., Chun, B., Cox, L., Jung, J., McDaniel, P. and Sheth, A. 2010. TaintDroid: An information-flow tracking system for realtime privacy monitoring on smartphones. OSDI.

FBI (Federal Bureau of Investigation). 2012. New e-scams & warnings. http://www.fbi.gov/scams-safety/e-scams (retrieved May 24, 2012).

Georgia Tech. October 18, 2011. Smartphones' accelerometer can track strokes on nearby keyboards. http://www.gatech.edu/newsroom/release.html?nid=71506 (retrieved June 21, 2012).

Govindarasu, M., Hahn, A. and Sauer, P. May 2012. Cyber-physical systems security for smart grid. Power Systems Engineering Research Center, publication 12-02.

Grace, N. 2012. FCC Advisory Committee adopts recommendations to minimize three major cyber threats, including an anit-bot code of conduct, IP route hijacking industry framework and secure DNS best practices. http://www.fcc.gov/document/csric-adopts-recs-minimize-three-major-cyber-threats (retrieved June 22, 2012).

Hewlett-Packard Development Company. March 2013. HP 2012 cyber risk report, white paper. http://www.hpenterprisesecurity.com/collateral/whitepaper/HP2012CyberRiskReport_0313.pdf (retrieved April 9, 2013).

Hulburt, G., Voas, J. and Miller, K. 2011. Mobile-app addiction: Threat to security? *IT Professional* 13:9–11.

IBM. September 2011. IBM X-Force 2011 mid-year trend and risk report. http://www-935.ibm.com/services/us/iss/xforce/trendreports/ (retrieved June 1, 2012).

Internet Crime Complaint Center. 2011. 2011 Internet crime report. http://www.ic3.gov/media/annualreport/2011_IC3Report.pdf (retrieved December 28, 2012).

Jansen, W. and Grance, T. December 2011. Guidelines on security and privacy in public cloud computing. Special publication 800-144. http://csrc.nist.gov/publications/nistpubs/800-144/SP800-144.pdf.

Kaplan, D. September 16, 2008. Google Docs flaw could allow others to see personal files. *SC Magazine.* http://www.scmagazine.com/Google-Docs-flaw-could-allow-others-to-see-personal-files/article/116703/?DCMP=EMC-SCUS_Newswire (retrieved June 1, 2012).

Laplante, P. and DeFranco, J. 2010. Another ode to paranoia. *IT Professional* 12:57–59.

Lipson, H. 2002. Tracking and tracing cyber-attacks: Technical challenges and global policy issues. Special report CMU/SEI-2002-SR-009.

Long, J. 2008. *No tech hacking: A guide to social engineering, dumpster diving, and shoulder surfing.* Burlington, MA: Syngress.

Luscombe, B. August 2011. 10 Questions for Kevin Mitnick. http://www.time.com/time/magazine/article/0,9171,2089344-1,00.html (retrieved May 18, 2012).

Mandiant. January 2013. APT1 exposing one of China's cyber espionage units. www.mandiant.com (retrieved April 10, 2013).

McAfee Labs. 2012. Threats report: First quarter 2012. http://www.mcafee.com/us/resources/reports/rp-quarterly-threat-q1-2012.pdf (retrieved June 22, 2012).

Mell, P. and Grance, T. September 2011. The NIST definition of cloud computing. Special publication 800-145. http://csrc.nist.gov/publications/nistpubs/800-145/SP800-145.pdf.

Mitnick, K. and Simon, W. 2002. *The art of deception.* New York: Wiley Publishing.

Mokey, N. July 19, 2010. Wallpaper apps swiped personal details off android Phones. *Digital Trends.* http://www.digitaltrends.com/mobile/wallpaper-apps-swiped-personal-details-off-android-phones/ (retrieved May 18, 2012).

Nakashima, E. 2010. More than 75,000 computer systems hacked in one of largest cyber attacks, security firm says. *Washington Post,* February 19.

Office of the Director of National Intelligence. October 2011. Foreign spies stealing US economic secrets in cyberspace—Report to Congress on foreign economic collection and industrial espionage. http://www.ncix.gov/publications/reports/fecie_all/Foreign_Economic_Collection_2011.pdf (retrieved May 24, 2012).

Rogers, D. July 8, 2012. How phone hacking worked and how to make sure you're not a victim. Nakedsecurity. http://nakedsecurity.sophos.com/2011/07/08/how-phone-hacking-worked/ (retrieved June 1, 2012).

Sonne, P. June 1, 2012. News corp. Faces wave of phone-hacking cases. *Wall Street Journal.* http://online.wsj.com/article/SB10001424052702303640104577440060134799828.html (retrieved June 1, 2012).

Turton, W., Mehrotra, K. June 2021. Hackers breached colonial pipeline using compromised password. https://www.bloomberg.com/news/articles/2021-06-04/hackers-breached-colonial-pipeline-using-compromised-password (retrieved March 6, 2022).

US Congress. February 2004. Annual report to Congress on foreign economic collection and industrial espionage—2003, NCIX 2004-1003. http://www.fas.org/irp/ops/ci/docs/2002.pdf (retrieved May 24, 2012).

US Department of Justice, Federal Bureau of Investigation. n.d. Business travel brochure. http://www.fbi.gov/about-us/investigate/counterintelligence/business-brochure (retrieved May 24, 2012).

Weiner, Z. 2012. Hacking. http://www.smbc-comics.com/ (retrieved February 20, 2012).

Wilson, C. n.d. 15-year-old admits hacking NASA computers. http://abcnews.go.com/Technology/story?id=99316&page=1 (retrieved June 17, 2012).

## NOTES

1 The first juvenile hacker to be incarcerated for computer crimes was 15 years old. He pled guilty and received a 6-month sentence in jail. He caused a 21-day interruption of NASA computers, invaded a Pentagon weapons computer system, and intercepted 3,300 e-mails and passwords (Wilson, ABC News).

2 Excerpt from Laplante and DeFranco (2010).

3 An APT is an attack where hackers infiltrate the corporate network and steal sensitive data over a long period of time. APTs will be addressed in Chapter 4.

# 2 Cyber Security

Distrust and caution are the parents of security.

*—Benjamin Franklin*

## 2.1 INTRODUCTION

And the winner is... Heartland Payment Systems for the single largest data breach in US history as of 2009. Since then, there have been many more data breaches. A few are shown in Table 2.1. Honestly it is debatable which breach gets the W for the most information leaks. Yahoo in December 2016 announced the incident that occurred in 2013 and leaked 3 billion accounts, and Alibaba leaked 1.1 billion pieces of user data in November 2019.

All security compromises are devastating no matter the size. Part of the recovery is to analyze and determine the cause in order to learn and improve security practices. Heartland is a business that provides payment transactions, which means it acts as an intermediary between merchants and the banks and clearly has a significant amount of personal customer data stored on its systems. The data that were stolen, as I am sure you have already guessed, were payment card data (130 million credit card numbers, expiration dates, and cardholder names). The Heartland Payment System's hacker, who also is responsible for the TJX breach in 2007, used an SQL (structured query language) injection that exploited the vulnerabilities in the database layer of the company's website with simple database commands. SQL injections are unfortunately not uncommon and won't go away! In December 2020, an SQL injection caused 8.3 million user accounts to be compromised at a company called Freepik (Madou 2021).

The scariest part of this data breach is that Heartland was fully compliant with the Payment Card Industry Data Security Standard (PCI-DSS) at the time of the intrusion. The takeaway from this incident is that "compliance with industry standards is no guarantee of security" (Vijayan 2010). Thus, the need to go beyond the standards is a must. It is a wake-up call for companies that feel they are secure by passing the PCI security audit. This is not to say that PCI and other standards are not good, but rather is simply pointing out that companies need to monitor their assets and points of entry continuously. For example, companies need to realize that they need to protect not only the most critical servers but also the servers that control things such as heating, venting, and air conditioning—"the ones that seem less important," said Peter Tippett, vice president of technology and innovation at Verizon Business (King 2009). If you ask a hacker, he or she will tell you that those "noncritical," possibly forgotten servers are the ticket inside your network. In sum, risk assessment needs to be proactive and continuous because the threats are continuously changing.

Whether you are a manager that needs to establish and implement an information security program or an engineer that wants to understand the information security program where you work, this chapter is a great place to start, as it is an overview

DOI: 10.1201/9781003245223-2

**TABLE 2.1**

**Large Data Breaches**

| | | |
|---|---|---|
| LinkedIn | April 2021 | 700 million accounts |
| Facebook | April 2020 | 530 million accounts |
| MGM Resorts | February 2020 | 10.7 million guests |
| EasyJet Airline | May 2020 | Over 9 million accounts |
| Door Dash | September 2019 | 4.9 million customer |

of the major components that are recommended to be part of any security program. First, let us define two terms: information security and cyber security. *Information security* is the process of protecting data against unauthorized access while ensuring its availability, privacy, and integrity. *Cyber security* is the body of technologies, processes, and practices designed to protect networks, computers, programs, and data from attack, damage, or unauthorized access; in other words, it implies more than the protection of data. Therefore, information security can really be considered a subset of cyber security. However, in reality, these terms are used interchangeably. We can conclude that the difference between information security and cyber security is that cyber security includes a few additional elements such as application security, network security, disaster recovery, and business continuity planning. Just in case there are any linguists reading this chapter, the history of the word "cyber" was explained by Ed Felten in his 2008 article, "What's the Cyber in Cyber-Security?" Essentially, it started with a Greek word that implies a boat operator; then Plato used it to mean "governance," and then, in the twentieth century, Norbert Wiener used "cybernetics" to refer to the robot controller. Finally, William Gibson in his novels about the future coined the word "cyberspace." Now it seems that the word "cyber" is put in front of anything and everything associated with the Internet.

## 2.2   INFORMATION SECURITY

Information security is a system consisting of many parts: software, hardware, data, people, procedures, and networks (Whitman and Mattord 2012). Each component of the system clearly has different security requirements, but they are all based on the CIA Security Triad Model (44 United States Code, Section 3542). The following are the official definitions of the security characteristics as well as the model (Figure 2.1):

> *Availability:* Ensuring reliable access to and use of information at all times
> *Confidentiality:* Preserving authorized **restrictions on access and disclosure**, including means for protecting personal privacy and proprietary information
> *Integrity:* Guarding against improper information modification or destruction, and **ensuring information nonrepudiation** (nondenial or proof of something) **and authenticity**

To ensure that these characteristics are integrated into a plan to secure the critical assets of the enterprise, the information security professional needs continuously to

**FIGURE 2.1** Security triad.

assess everything and anything that would affect the integrity, availability, and confidentially of the information (e.g., unauthorized access, destruction, modification, natural disasters, power failure) because, as we mentioned, the threats and technology keep changing.

This is clearly a massive topic that has been covered in many books and articles. The goal of this chapter is not to turn someone into an information or cyber security professional but rather to provide a foundation to understand the main topics of information and cyber security. Let us begin with the established information, concepts, and activities that make up information security—in other words, what is the common body of knowledge (CBK) for information security.

Researchers Theoharidou and Gritzalis (2007) created a classification scheme outlining the ten basic domains for the information security profession based on their review and analysis of curricula, courses, university programs, and industry needs.

1. Security architectures and models
2. Access control systems and methodologies
3. Cryptography
4. Network and telecommunications and security
5. Operating system security
6. Program and application security
7. Database security
8. Business and management of information system security
9. Physical security and critical infrastructure protection
10. Social, ethical, and legal consideration

Not surprisingly, most of these domains overlap with the CBK of the CISSP (certified information systems security professional) certification (Table 2.2).

The remainder of this chapter will be dedicated to an overview of each element in the information security CBK.

## 2.3 SECURITY ARCHITECTURE

Security architecture is essentially a description or plan of how the *security controls* are related to information systems. The controls are what will help maintain the triad (integrity, availability, and confidentiality). In their book, *Principles of Information*

**TABLE 2.2**
**Information Security CBK**

| CISSP Domains | Theoharidou Domains |
|---|---|
| Access control | Access control systems and methodologies |
| Telecommunications and network security | Network and telecommunications security |
| Software development security | Program and application security |
| Cryptography | Cryptography |
| Security architecture and design | Security architecture and models |
| Business continuity and disaster recovery | Business and management of information systems security |
| Legal, regulations, investigations, and compliance | Social, ethical, and legal consideration |
| Physical (environmental) security | Physical security and critical infrastructure protection |
| Information security governance and risk management | Operating systems security |
| Operations security | Database security |

*Security* (2012), Whitman and Mattord describe the many elements that complete a security architecture design. A few of the components—spheres of security, levels of control, defense in depth, and security perimeter—will be summarized in this section.

A version of their *sphere of security* is shown in Figure 2.2. The sphere illustrates how information is accessed and needs to be protected. Listed next to each layer name are the protection elements. For example, "education/training" and "policy and law" are the protection between people and information. The protections between the Internet and networks are access controls, host intrusion, detection, and prevention systems (IDPS), firewalls, monitoring systems (sniffers), redundancy (implementing backup security), security planning (business continuity, disaster recovery, and incident response), and policy and law. All of these protection mechanisms will be discussed in upcoming sections.

There are three types of *levels of control:* managerial, operational, and technical. At the managerial control level, the security strategic plan is determined as well as risk management, security control reviews, and setting the guidelines to maintain any compliance required. At the operational control level, disaster recovery and incident response are planned. Also addressed at this level are personnel and physical security. The operational controls are integrated into specific business functions and the technical control level covers the components that protect the information assets.

*Defense in depth* is the layered implementation of security where the layers are policy (e.g., preparing the organization to handle attacks), training and education (e.g., defending against social engineering tactics), and technology (multilayered: firewalls, proxy servers, and access controls).

And, finally, the *security perimeter* defines the boundary between the organization's security and external entities. This is at the network level and the physical level (e.g., entrance to the building).

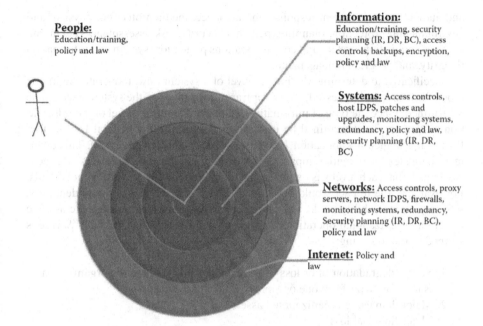

**People:**
Education/training,
policy and law

**Information:**
Education/training, security
planning (IR, DR, BC), access
controls, backups, encryption,
policy and law

**Systems:** Access controls,
host IDPS, patches and
upgrades, monitoring systems,
redundancy, policy and law,
security planning (IR, DR,
BC)

**Networks:** Access controls, proxy
servers, network IDPS, firewalls,
monitoring systems, redundancy,
Security planning (IR, DR, BC),
policy and law

**Internet:** Policy and
law

**FIGURE 2.2**  Spheres of security. (Adapted from Whitman, M., and Mattord, H., 2012, *Principles of Information Security*. Stamford, CT: Cengage Learning, Course Technology.)

## 2.4   ACCESS CONTROLS

Access control refers to a security feature that manages the interaction with a resource. The resource can be either technology or a room containing the technology (physical security). In this section, we will focus on access control with technology and discuss physical later in this chapter.

Access control is critical to minimizing system vulnerabilities. It restricts not only who or what has access to the resource but also the type of access that is permitted. Here is access control explained in five "easy" steps (Gregory 2010):

1. Reliably identify the user/system
2. Find out what resource the user/system wishes to access
3. Determine if the user/system has permission to access the resource
4. Permit or deny access
5. Repeat

For an access control plan that is slightly more detailed than this one, let us refer to NIST (National Institute of Standards and Technology). They have documented security control baselines that are recommended to be employed depending on the impact level (information dependent) of the information system (low, moderate, high). In addition to access control, NIST also recommends controls for the following categories: awareness and training, audit and accountability, security assessment and authorization, configuration management, contingency planning, identification and authentication, incident response, maintenance, media protection, physical and environmental protection, planning, personal security, risk assessment, system and services acquisition, system and communications protection, system and information integrity, and program management.

Specifically, to determine the impact level of a system (low, moderate, high), the type of information processed, stored, or transmitted by or on the system needs to be determined. Once the type of information is known, the impact level of the information system can be determined by using the guidelines in the Federal Information Processing Standard Publication 199 (Evans, Bond, and Bement 2004). This document indicates the potential impact (level) by defining the adverse effects of a security breach that each level has on the organization's operations, assets, or individuals. For example, the impact would be defined as HIGH if the "loss of confidentiality, integrity, or availability could be expected to have a severe or catastrophic adverse effect on organizational operations, organizational assets, or individuals." *Severe* is defined as the following:

1. Severe degradation in or loss of mission capability where the organization is not able to perform one or more of its primary functions
2. Major damage to organizational assets
3. Major financial loss
4. Severe harm to individuals (loss of life or life-threatening injuries)

An impact level is determined separately for the *confidentiality*, the *integrity*, and the *availability* of the information on the system. The final security category (SC) will then be determined by the highest impact level of the *confidentiality*, *integrity*, or *availability* of the system. An example of an SC expression of a high-impact information system is as follows:

SC = {(confidentiality, low), (integrity, high), (availability, moderate)}

For the impact to be considered *low*, all three categories need to be *low*; to be considered moderate, there must be only low and moderate levels indicated. Thus, in the preceding example, the impact level is *high* since an **integrity** loss of the information on that system was evaluated to be *high* impact. Once the SC is determined, the security controls can be selected, tailored, and supplemented if needed (Locke and Gallagher 2009).

Each security control listed in each category has an associated priority code. The priority code indicates the recommended sequencing of implementation. For example, ALL controls with the highest priority are implemented first (if that control is indicated for the system's impact level), and then implementation of the controls in the next priority level would be implemented, and so on.

The control baselines for each impact level are listed in the NIST special publication 800–53, "Recommended Security Controls for Federal Information Systems and Organizations." Following is a simplified view of the *priority 1* recommendations for the **access control** baseline. Note that a control may be required for multiple impact levels but that level may require additional "control enhancements" (added functionality) to increase the strength of that particular control. The control enhancements are not specified in the following table:

| Control Name | Impact Level | | |
|---|---|---|---|
| | Low | Moderate | High |
| Develop, distribute, and maintain access control policy and procedures | ✓ | ✓ | ✓ |
| Manage IS user accounts | ✓ | ✓ | ✓ |
| Enforce access to system according to policy | ✓ | ✓ | ✓ |
| Control the information flow between and within systems | Not selected | ✓ | ✓ |
| Least privilege (allowing enough access to do one's job) | Not selected | ✓ | ✓ |
| Display privacy and security notifications to users | ✓ | ✓ | ✓ |
| Identify permitted actions without identification or authentication | ✓ | ✓ | ✓ |
| Document methods of remote access | ✓ | ✓ | ✓ |
| Establish usage restrictions and guidance for the implementation of wireless access | ✓ | ✓ | ✓ |
| Establish access control for mobile devices | ✓ | ✓ | ✓ |
| Establish the terms and conditions when using external information systems | ✓ | ✓ | ✓ |

Of course, prior to any access or the controlling of access, the system needs first to identify the user (user ID). The NIST suggests the following considerations for user identification, authentication, and passwords (Swanson and Guttman 1996).

**Identification** means to *require identification* of users and *correlate actions to users* by having the system maintain the IDs of active users and linking those users to identified authorized actions. The *maintenance of user IDs* occurs by deleting former and inactive users as well as adding new users.

**Authentication:** The system also needs to validate the identification claim. Possible means to authenticate are something the user *knows* (e.g., password, personal identification number, or cryptographic key), something the user *possesses* (e.g., smartcard or USB token), or something the user *is* (e.g., a biometric such as a fingerprint, voice recognition, iris scan, facial scan, handwriting, or voice recognition). Authentication can become tricky, however, due to the bring-your-own-device (BYOD) model. More and more companies are dealing with the use of employee-owned devices connected to the corporate network. For example, what if that device, which probably contains sensitive data or documents, is lost or stolen? One solution is the Tactivo smart casing shown in Figure 2.3. The sensitive information can only be accessed via a smartcard or fingerprint or both depending on the security policy.

The biometric has the greatest advantage since a user ID, password, and token can be more easily impersonated than the biometric (Gregory 2010). The downsides are the cost of maintaining biometric implementation, gradual or sudden changes in user characteristics, and false readings.

Other safeguards suggested by NIST for authentication are requiring users to authenticate, restricting access to authentication data, securing transmission of

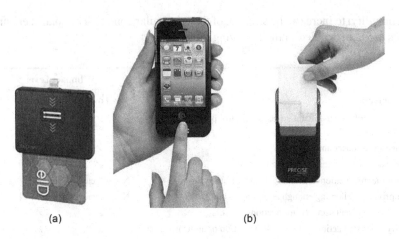

(a)                                                    (b)

**FIGURE 2.3**  Tactivo finger swipe and smartcard readers from PreCise Biometrics (http://www.precisebiometrics.com/).

authentication data, limiting log-on attempts, securing authentication as it is entered (suppress display), and administering data properly (disable lost/stolen passwords or token and monitor systems looking for stolen or shared accounts).

Note that passwords should have required attributes (minimum length and special characters) and should be changed frequently. Users should also be trained not to "give it away" (e.g., easily guessed, posting it on their computer, telling a friend).

In addition to identification and authentication, access control needs to include authorization and accountability (Whitman and Mattord 2012).

**Authorization** is matching the authenticated entity to a list of access and control levels. This can be accomplished individually or as a group. An access control list can be used. An access control list is a list of users who have been given permission to use a particular resource; in addition to the type of access they have been permitted.

**Accountability** is using logs and journals to record the use or attempted use of a particular resource.

There are also many access control technologies. Two popular technologies are *Kerberos* and *Active Directory*. Kerberos is a network authentication protocol one can obtain from Massachusetts Institute of Technology (MIT) (http://web.mit.edu/kerberos/). Kerberos uses strong cryptography to connect a client and server to each other and is able to assist in encrypting the communication between the two over a nonsecure network (the hosts must be trusted, however). The premise of Kerberos is that it not only prevents outside attacks but can also minimize the attacks that occur from within the firewall. This is done by not sending passwords over the network in cleartext. Instead, it actually sends a "key" to the password. Active directory (based off the lightweight directory access protocol) is a domain controller that maintains user account and login information. Active Directory is used by Windows to keep track of the access and security rights assigned to each user.

## 2.5   CRYPTOGRAPHY

Cryptography is used to conceal and disguise information. Cryptography uses an encryption algorithm and a key to change data until they are unrecognizable. In other words, it uses an algorithm to scramble the data so that only a person with the proper key can unscramble them. Cryptography not only provides confidentiality and integrity of the data, but it also can provide nonrepudiation (Conrad 2011). Nonrepudiation means the authenticity of the data or information cannot be challenged.

Cryptography is not a new concept. It was used over 2,000 years ago in the days of Julius Caesar for military purposes where he communicated with his troops using a shift cipher (aka Caesar cipher or substitution cipher). In a shift cipher, each plaintext letter is replaced with another letter a certain number of places further down in the alphabet. The number of places is called the *key*.

An Example

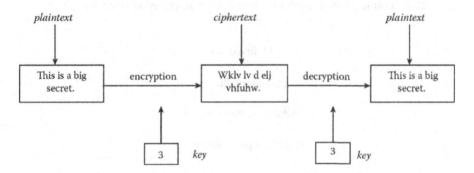

In modern times, obviously, the encryption needs to be more complex but the bottom line is that it depends on the needs of the organization and the sensitivity of the data. Here are a few, more complex cipher methods used to encrypt plaintext (Whitman and Mattord 2012):

*Transposition cipher:* The values with a block are rearranged to create the ciphertext. The rearrangement depends on the key pattern. For example, if the key pattern states:

| Text position | 1 | 2 | 3 | 4 | 5 | 6 | 7 | 8 |
|---|---|---|---|---|---|---|---|---|
| Moves to position | 3 | 1 | 4 | 5 | 7 | 8 | 2 | 6 |

Use the key pattern to rearrange the plaintext to ciphertext:

| Text position | 8 | 7 | 6 | 5 | 4 | 3 | 2 | 1 |
|---|---|---|---|---|---|---|---|---|
| Plaintext | S | T | R | E | N | G | T | H |
| Ciphertext | R | E | S | N | G | H | T | T |

*Exclusive OR (XOR):* Two bits are compared. If both are the same, the result is 0. If they are different, the result is a 1. To perform the XOR cipher method, the XOR is performed on a key value with the value being encrypted—for example:

| | | |
|---|---|---|
| Plaintext | A | 01000001 |
| Key | V | 01010110 |
| Ciphertext | A XOR V | 00010111 |

*One-time pads (aka Vernam cipher):* This cipher, invented by Gilbert Vernam while he was working at AT&T, adds a random set of characters (only used one time) to a block of plaintext (of the same length). Each character of the plaintext is turned into a number and a pad value is added to it. The resulting sum is converted to ciphertext. If the sum of the two values is above 26, then 26 is subtracted from the total. Here is an example using one letter:

Plaintext = E

Plaintext value = 5

One-time pad text = J

One-time pad value = 10

Since the sum of plaintext and the pad is 15 (thus less than 26), it is converted to the ciphertext letter O.

*Book cipher:* The ciphertext contains a list of codes that represent the page number, line number, and word number of the plaintext in a particular book. The encoded message may look like this (50, 20, 5; 75, 30, 3; 300, 4, 7). Thus, on page 50, line 20, word number 5 is the first word in the message. The second word in the message is on page 75, line 30, word number 3, and so on.

### 2.5.1   TYPES OF CRYPTOGRAPHY OR CRYPTOGRAPHIC ALGORITHMS

There are essentially three types of encryption: symmetric, asymmetric, and hashing (Conrad 2011):

*Symmetric key encryption:* Using the same key to encrypt and to decrypt. This encryption technique came first. The disadvantage to this type of encryption is that it is difficult to distribute the key securely. If the wrong person gets the key, you are out of luck because anyone who has the key can decrypt your message. There are many different types of symmetric encryption. The most widely known, developed by IBM, is data encryption standard (DES). The successors to DES are triple DES (3DES) and advanced encryption standard.

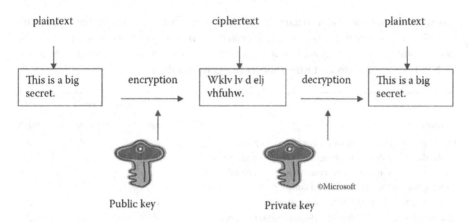

**FIGURE 2.4** Asymmetric encryption example.

This type of encryption can be used if the two computers that are communicating are known so that the key can be installed on each computer.

*Asymmetric key encryption:* Also known as public-key cryptography, asymmetric key encryption uses one key to encrypt and a different key to decrypt. The encryption key is a public key and anyone with a copy of the key can encrypt information that only you can read with the private key for decryption. This type of encryption solves the problem of symmetric key encryption, that the key can be easily accessed by an attacker. The private key is only known by your computer, while the public key is known by anyone who wants to communicate securely with your computer. Even if the message is intercepted in transit, you can only read it with the private key. An example can be found in Figure 2.4.

*Hash functions:* No key is used for hashing. Hash functions are mostly used to confirm data integrity (that the data have not changed). There are many hash algorithms used. One of the widely used hash algorithms is MD5. A good algorithm is determined by having a limited number (e.g., one in one billion) of collisions (two distinct files resulting in the same hash value). When the hash algorithm is performed on the plaintext, a hash value is created. Therefore, if the plaintext changes, so will the hash value.

## 2.6 NETWORK AND TELECOMMUNICATIONS SECURITY

The *telecommunications* and *network security* knowledge domain includes a vast amount of concepts and technology needed to protect the security triad (confidentiality, integrity, and availability). These include fundamental network concepts such as network structures, transmission methods, and transport formats in order to provide authentication for transmissions over private and public communications networks and media. Clearly, someone working in the cyber security area would need a deep understanding of those fundamental concepts to know what he or she is protecting. However, that topic deserves its own book and thus is out of the scope of this section. The focus and goal of this section is for the reader to gain an understanding about the controls that need to be considered to secure the *networks* and to protect the *transmission* of data over a network (*telecommunications*). Following is a simplified view of the *priority 1* recommendations by NIST for the **system and information integrity** baseline:

| Control Name | Impact Level | | |
| --- | --- | --- | --- |
| | Low | Moderate | High |
| Develop system and information integrity policy and procedures that define, facilitate, and implement the roles, responsibilities, management commitment, and coordination among organizational entities and compliance | ✓ | ✓ | ✓ |
| Test, correct, and report information system flaws | ✓ | ✓ | ✓ |
| Malicious code protection at all information system entry points and any device connected to the network | ✓ | ✓ | ✓ |
| Information system monitoring to identify unauthorized use and any attacks | Not selected | ✓ | ✓ |
| Continually receive security alerts, advisories, and directives from a designated external organization | ✓ | ✓ | ✓ |
| Verify the correct operation of security functions | Not selected | Not selected | ✓ |
| The information system monitors software and information integrity | Not selected | ✓ | ✓ |
| Spam protection at all information system entry points and any device connected to the network | Not selected | ✓ | ✓ |
| The system validates information input | Not selected | ✓ | ✓ |

There are other control baselines that could also be applied to the security of network and telecommunications. It would be best to review the NIST special publication 800–53 to determine which controls would be most effective for your situation.

## 2.7   OPERATING SYSTEM SECURITY

The topic of operating system (OS) security overlaps with many of the topics in this chapter. The most essential security mechanism for an OS is *access control*, which was covered in Section 2.4. The OS actually facilitates the enforcement mechanism of the access control (e.g., which users or systems can perform which operations using which resources). As mentioned previously, the type of access control is determined by the protection needed. The amount of protection is based on the system's impact level (low, moderate, high).

Jaeger (2008) describes three guarantees of a secure operating system:

1. *Complete mediation:* All security-sensitive operations are facilitated.
2. *Tamperproof:* The access enforcement cannot be modified by untrusted processes.
3. *Verifiable:* The system should be testable to demonstrate that security goals are being met.

Many companies have mail servers; hence, reviewing the NIST-recommended checklist for securing a mail server operating system is a great example of an OS security task (Tracy et al. 2007):

| Category | Action | Completed |
|---|---|---|
| *Patch and upgrade* | | ✓ |
| | Create and implement a patching process | |
| | ID, test, install patches and upgrades to OS | |
| *Remove or disable unnecessary services and applications* | | |
| | Remove or disable unnecessary services and applications | |
| | Use separate hosts for other services (web, directories, etc.) | |
| *Configure operating system user authentication* | | |
| | Remove or disable unneeded default accounts and groups | |
| | Disable noninteractive accounts | |
| | Create the user groups for the particular computer | |
| | Create the user accounts for the particular computer | |
| | Create an effective password policy (e.g., length, complexity) and set accounts appropriately | |
| | Configure computers to prevent password guessing | |
| | Strengthen authentication by installing and configuring other security mechanisms | |
| *Configure resource controls appropriately* | | |
| | Set access controls for resources (e.g., files, directories, devices) | |
| | Limit privileges to authorized administrators for most system-related tools | |
| *Install and configure additional security controls* | | |
| | Install and configure software to provide additional controls not available in the OS | |
| *Test the security of the OS* | | |
| | Test OS after initial installation for vulnerabilities | |
| | Periodically test OS for new vulnerabilities | |

## 2.8  SOFTWARE DEVELOPMENT SECURITY

To develop and maintain software free from security problems is no easy task. It is one of the many nonfunctional requirements a software engineer needs to design into software (e.g., usability, maintainability, scalability, availability, extensibility, security, and portability). But, in this era, special attention needs to be paid to the security requirement. This can be achieved by building security into applications during the development process (Khan and Zulkernine 2008). However, two of the difficulties software developers face are the lack of application security knowledge and schedule pressures (Payne 2010). The eight-step process developed by Talukder et al. (2009) to elicit both functional and nonfunctional security requirements can be part of the solution where the security issues of the application are analyzed upfront. Following is the summarized version of the eight steps:

1. *Functional requirements:* Capture requirements using Unified Modeling Language (UML) analysis artifacts.
2. *Identification of assets:* Identify the critical assets of the organization and categorize them by their perceived value and loss impact.
3. *Security requirements:* Determine possibilities (diagram a misuse case; see the example in Figure 2.5) for attacks (e.g., denial of service [DOS], data tampering) and tampering with the data characteristics (e.g., confidentiality, integrity, and availability).
4. *Threat and attack tree:* Analyze each misuse case and determine the threat path.
5. *Rating of risks:* Assign values to each threat/risk to determine the highest risk.
6. *Decision on in vivo versus in vitro:* Determine which threats need to be addressed within the application (in vivo). This is done by comparing the threats to the assets.
7. *Nonfunctional to functional requirement:* Move threats that are in vivo to the functional requirements.
8. *Iterate:* Revisit and refine requirements.

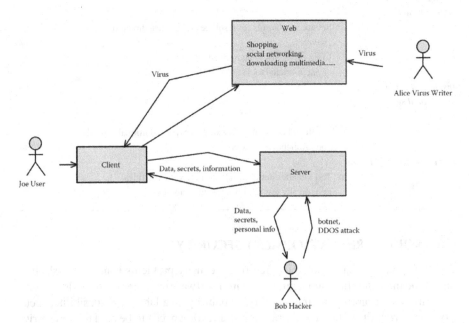

**FIGURE 2.5**   Misuse case of risk assessment.

No matter how diligent our software developers become in incorporating security in their software applications, we can guarantee that hackers will remain persistent in their efforts to find new ways to exploit the vulnerabilities in software. In an effort to educate software developers on common software weaknesses, the SANS Institute and the MITRE Corporation (a not-for-profit organization) collaborated with many top security experts to develop the list of the "Top 25 Most Dangerous Software Errors" (Martin et al. 2011). Over 20 organizations provided input on this list. These errors are ranked based on their evaluation of the prevalence, importance, and likelihood of each weakness. SQL injections are the top software error—which is no surprise.

| Score | ID | Name |
| --- | --- | --- |
| 93.8 | CWE-89 | Improper neutralization of special elements used in an SQL command ("SQL injection") |
| 83.3 | CWE-78 | Improper neutralization of special elements used in an OS command ("OS command injection") |
| 79.0 | CWE-120 | Buffer copy without checking size of input ("classic buffer overflow") |
| 77.7 | CWE-79 | Improper neutralization of input during web page generation ("cross-site scripting") |
| 76.9 | CWE-306 | Missing authentication for critical function |
| 76.8 | CWE-862 | Missing authorization |
| 75.0 | CWE-798 | Use of hard-coded credentials |
| 75.0 | CWE-311 | Missing encryption of sensitive data |
| 74.0 | CWE-434 | Unrestricted upload of file with dangerous type |
| 73.8 | CWE-807 | Reliance on untrusted inputs in a security decision |
| 73.1 | CWE-250 | Execution with unnecessary privileges |
| 70.1 | CWE-352 | Cross-site request forgery (CSRF) |
| 69.3 | CWE-22 | Improper limitation of a pathname to a restricted directory ("path traversal") |
| 68.5 | CWE-494 | Download of code without integrity check |
| 67.8 | CWE-863 | Incorrect authorization |
| 66.0 | CWE-829 | Inclusion of functionality from untrusted control sphere |
| 65.5 | CWE-732 | Incorrect permission assignment for critical resource |
| 64.6 | CWE-676 | Use of potentially dangerous function |
| 64.1 | CWE-327 | Use of a broken or risky cryptographic algorithm |
| 62.4 | CWE-131 | Incorrect calculation of buffer size |
| 61.5 | CWE-307 | Improper restriction of excessive authentication attempts |
| 61.1 | CWE-601 | URL redirection to untrusted site ("open redirect") |
| 61.0 | CWE-134 | Uncontrolled format string |
| 60.3 | CWE-190 | Integer overflow or wraparound |
| 59.9 | CWE-759 | Use of a one-way hash without a salt |

MITRE maintains the common weakness enumeration (CWE). For details of each software weakness, please visit cwe.mitre.org.

Lastly, software developers have many different methodologies and processes they prefer when they analyze, design, and implement their software. No matter which process they use, a few additional measures can be taken to help in the process of building secure software (Gregory 2010):

1. *Provide source code access control:* The only people with access to the source code should be authorized developers and even fewer people on the development team should be able to modify the source code.
2. *Provide protection of software development tools:* Reducing access and thus modification of the development tools and libraries will reduce the possibility of introducing vulnerabilities through tampering with the tools.
3. *Provide protection of the software development systems:* The servers used to house the source code repositories should be protected at the same level as the application servers. The hackers are looking for any weak link they can find.

## 2.9 DATABASE SECURITY

Databases are probably the most essential system a company owns. If a hacker is able to communicate with the company database, he now has access to business-critical data. The type of information that typical database systems house (customer, financial, and company details) can put the future of the business in question if that information gets into the wrong hands. Thus, engineers need to be aware of vulnerabilities and threats, attacks on databases, secure architectures, security models for the next generation of databases, security mechanisms in databases, security database design, and the attributes of database security. This is quite a challenge even for the most experienced, as we see in the news about stolen data from high-profile companies. In the previous section we showed the "Top 25 Most Dangerous Software Errors," where the SQL injection is number 1!

Database security is complex because it is not only about protecting against intentional theft or accidental exposure, but also about compromise of data by employee disgruntlement or ignorance. Data loss prevention (DLP) strategies include:

- Data inventory
- Data classification
- Data metric collection
- Policy development for data creation, use, storage, transmission, and disposal
- Tools to monitor data at rest, in use, and in transit

The preceding can be accomplished by using typical network and security tools such as network analysis software, application firewalls, and intrusion detection and prevention systems.

But how do we mitigate the number 1 problem of SQL injections where criminals are modifying database queries to steal, corrupt, and change private data?

The MITRE and SANS corporations suggest many specific prevention measures to mitigate SQL injection, including architecture and design decisions, implementation, and operation of the database in their CWE document cited in the previous section (http://cwe.mitre.org/top25/).

## 2.10   INTERNET OF THINGS SECURITY

Bad actors are always more agile than everyone else when it comes to new technology. Their process is simple: (1) find the vulnerability of the new technology; (2) determine the easiest way to compromise it; and (3) determine how to monetize it. For the consumer, defending against the bad actor can be simple, however, the consumer usually doesn't think or know about system vulnerability or how to address it. Generally, this difference between the bad actor and consumer, when it comes to new technology, is also apparent by the time difference between the bad actor taking advantage of a new technology and the time it takes to implement a defense against it. We will name this difference the *agility gap*. The agility gap can be explained as: a new technology appears—the bad actors are quick to find a vulnerability while the rest of us simply enjoy the smart plugs and beer ordering refrigerator until a problem is noticed.

The agility gap is especially apparent with Internet of Things (IoT) devices. In reality, IoT is becoming more ubiquitous every day as this technology is present in critical infrastructure, the educational system, our cities, healthcare, and homes. The problem we will highlight in this section is IoT technology in the home such as smart plugs (timer), lighting, security cameras, and self-ordering refrigerators. Although effective technology, as it simplifies many homeowner tasks, IoT also adds many entry points into our home network making it an easy target for a bad actor. One of the reasons for this is the availability of default login IDs and passwords on the Internet. A bad actor could simply download this default password list from the Internet to access the routers they are targeting. This is an easy attack because most homeowners do not change the manufacturer default password on IoT technology used in the home.

The attack method of using default user ID and passwords found on the Internet has been around for a long time and can be addressed with one of the first security principles: *when implementing new equipment, change the manufacturer default user ID and password.*

The problem is that the average person setting up an IoT device is not a security expert. Manufacturers make device setup user-friendly. For example, many devices come with a QR code making the setup of this device in the home very simple. So simple, that the average user doesn't know how to change the default password.

NIST provides guidelines for manufacturers such as in NIST in SP 800–53. This special publication recommends changing default passwords as a best practice for hardware implementations such as networking equipment like routers and switches or things like servers and services running on them (NIST 2020). Getting this message out to the average user should be a responsibility of the manufacturer and therefore should be integrated into the device.

NIST in SP 800–63b, also documented password recommendations that although intended for federal systems, the community at large follows these best practices (Grassi et al. 2017). Some of the recommendations include the following: *using*

*passphrases rather than passwords, not enforcing password resets, screen pass-*
*words for commonly used passwords, allow password pasting, limit the number of*
*failed password attempts, implement 2-factor authentication, salt and hash pass-*
*words,* and *enable "show password" while typing.*

The average user expects the manufacturer to bake the security into the device
such as managing updates and providing a secure password. The problem is that the
complexity of these systems is where the failure lies. For example, there may be six
or seven components in a system—six may be secure but the weakest link, the sev-
enth component, is the problem.

This can be described by thinking about third-party risk—the weakest link is not
the big company as they likely have an effective security program in place. Most of
the time third parties connected to the big company are the security issue as they
would be considered the weakest link. Therefore, the bad actors are trying to get
through to the big company through their third-party vendors.

IoT devices are similar to this third-party risk example. A typical user has dozens
of devices, all with an unchanged default password, that they are not managing nor
are they looking for software updates. Since the average home network does not use
enterprise level routers, the homeowner depends on the manufacturer to ensure the
security of the devices they are connecting to their homes.

Why should the average user care? Because their device could be part of a distributed denial of service (DDoS) Botnet. A *botnet* is a group of computers that
have been compromised by malware and become controlled by a malicious actor.
A *DDoS botnet* is when that group of computers overloads a server with web traffic. A typical user will not notice the degradation on the network. On the other
side, if there are 500K endpoints that are attacking a network, that is too many to
trace and are probably unsecured home routers. Thus, the homeowner will not be
informed.

Currently, ransomware is the number one cyber threat because it is very productive. However, the typical compromise using IoT is a DDoS botnet attack, and
they are growing in number and frequency. This is due to device manufacturers
sending out products that are not properly secured. These botnets are composed of

> IoT devices that have non-existent or inadequate passwords, inability to patch exploit-
> able firmware, or holes in the authentication and data transfer ecosystem. Automated
> attacks on known vulnerabilities have granted cyber criminals extensive ability to
> quickly assemble or grow a botnet.
>
> *(Ikeda 2020)*

It is important to be aware that IoT can "amplify the impact" of a ransomware
attack as it widens the attack surface. Trend Micro (2021) recommends securing
possible entry points: *update and patch, employ secure authentication strategies,*
*enforce the principle of least privilege, regularly back up files, ensure strong net-*
*work protection, monitor network traffic, prioritize security over connectivity,* and
*advocate shared responsibility over the IoT.*

Even though a DDoS may be the largest concern for IoT, ransomware is also an
issue. It was demonstrated at a Def Con hacking conference demonstrated an attack

using ransomware on a smart thermostat hundreds of miles away which forced a bitcoin payment for the homeowner could regain control of the appliance (Dickson 2016). This is most frightening for IoT medical devices.

Earlier in this section several strategies to prevent attacks were mentioned such as changing default passwords. In the recent Whitehouse memorandum (Young 2022), it was recommended to move to a zero-trust model which includes five areas: Identity, Devices, Networks, Applications and Workloads, and Data. A vision and actions for each area are also provided in this memorandum. For example, for the identity area, agencies must enforce strong multi-factor authentication. This is further explained with being enforced at the application layer, must be phishing-resistant, and password policies must not require special characters or regular rotation.

## 2.11 BUSINESS CONTINUITY AND DISASTER RECOVERY

Business continuity and disaster recovery (BC/DR) planning includes both response and recovery plans as well as restoration activities if disaster occurs. Disasters can be anything from a virus to a hurricane or cyber-attack. *Disaster recovery* includes a process that recovers critical information if a disaster happens to vital systems. *Business continuity* includes how the business will continue if the critical information is lost. An easy way to differentiate the two is that the BC is the process your company will follow in order to keep making money if disaster occurs. While the BC is in the works, the company can be performing DR, where the critical systems and information are recovered. Hence, one can appreciate why they both should be included in the same plan.

The essential pieces of the BC/DR plan will encompass how the immediate business needs can be met to resume operations. Generally, a plan would include the following (Dey 2011):

- Create a BC/DR team and define and assign roles to each team member.
- Conduct risk assessment to determine priorities and requirements.
- Create a plan including budget requirements.
- Create policies and procedures to be approved by management.
- Begin to implement the project (procure necessary resources).
- Arrange alternate network links and determine facility requirements (cold site, warm site, hot site). For example, the hot site would include the most resources (a fully configured computer facility).
- Achieve and maintain compliance, regulations, and best practices as they apply to your business.
- Link the BC/DR plan with the organization's change management process so that changes in the business process would be included in the BC/DR plan.
- Routinely test the plan.
- Review and iterate the BC/DR framework.

## 2.12 PHYSICAL SECURITY

Securing the physical environment refers to addressing the security of the facility both internally and at the perimeter. This is defined as critical infrastructure protection—in other words, sections of the facility that provide protection and support of the critical information systems. Following is a simplified view of the *priority 1* recommendations by NIST for the **physical and environment protection** baseline:

| | Impact Level | | |
| --- | --- | --- | --- |
| Control Name | Low | Moderate | High |
| Develop, distribute, and maintain physical and environmental protection policy and procedures | ✓ | ✓ | ✓ |
| Keep a current list of personnel authorized to access the facility | ✓ | ✓ | ✓ |
| Manage physical access control of the facility where information systems reside using devices and/or guards | ✓ | ✓ | ✓ |
| Access control for transmission medium within the facilities | | ✓ | ✓ |
| Access control for output devices (e.g., monitors, printers, audio devices) | | ✓ | ✓ |
| Monitor physical access | ✓ | ✓ | ✓ |
| Visitor control (authenticate individuals) | ✓ | ✓ | ✓ |
| Protect power equipment and cabling from damage | | ✓ | ✓ |
| Provide an emergency shutoff capability | | ✓ | ✓ |
| Provide short-term emergency power supply to shut down information systems properly in the event of primary power loss | | ✓ | ✓ |
| Provide emergency lighting in the event of a power outage | ✓ | ✓ | ✓ |
| Employ fire suppression and detection devices for information systems | ✓ | ✓ | ✓ |
| Implement temperature and humidity controls | ✓ | ✓ | ✓ |
| Provide water damage protection for information systems | ✓ | ✓ | ✓ |
| Monitor delivery area access | ✓ | ✓ | ✓ |
| Employ an alternate work site in the event of a security incident | | ✓ | ✓ |

## 2.13 LEGAL, REGULATIONS, COMPLIANCE, AND INVESTIGATIONS

In addition to technical experience, the cyber security professional needs to be familiar with the legal system and the many types of laws, regulations, and compliance as they pertain to information. This knowledge will help one to understand how to handle things like intellectual property and individual privacy, as well as how to conduct an investigation and handle evidence legally. These topics will be discussed in further detail in Chapter 6 but, for now, here are a few of the many laws and regulations with which a security professional should be acquainted:

- *Provide Appropriate Tools Required to Intercept and Obstruct Terrorism (PATRIOT) Act:* This act allows law enforcement to conduct electronic surveillance to investigate terrorists.

- *Privacy Act of 1974:* This act requires consent of a citizen for a person or agency to send information on that citizen.
- *Fourth Amendment:* This amendment to the Constitution states that the right

> to be secure in their persons, houses, papers, and effects, against unreasonable searches and seizures, shall not be violated, and no warrants shall issue, but upon probable cause, supported by Oath or affirmation, and particularly describing the place to be searched and the persons or things to be seized.

- *Health Insurance Portability and Accountability Act (HIPAA) of 1996:* This act specifies safeguards (administrative, physical, and technical) to ensure the confidentiality, integrity, and availability of electronic health information.

## 2.14 OPERATIONS SECURITY

Along with the other topics in this chapter, operations security also deals with a range of activities to minimize the risk of threats and vulnerabilities to the critical information of an organization. Operations security primarily entails identifying the information that needs protection and determining the most effective way to protect the information.

Understanding the key security operations concepts is a priority. An example is implementing access control mechanisms (discussed in Section 2.4) such as the concept of "need to know" where an individual only has access if the information is needed to perform his or her job. Another way to minimize risk is to implement "separation of duties," where no individual has control over an entire process, thus reducing the possibility of fraud. For example, changing a firewall rule should be performed by two people. The banking industry uses this concept in almost all of its processes; for example, no one person has the full combination to the bank vault. Thus, to open the vault you need two people, each knowing one-half of the combination. Other security operations concepts are rotating jobs (illegal activity would not likely be on the mind of a person being rotated out of a job) and monitoring special privileges to any type of system administrator (database, application, network), as he or she has the opportunity to corrupt data with privileged access.

Resource protection and incident response should also be considered as part of the operations security plan. The resources are physical entities such as hardware (e.g., equipment life cycle, cabling, and wireless networks) and software (e.g., licensing, access control, and source code protection). Incident response refers to a plan in place in the event that interruption of normal operations occurs. This plan would include how to detect and determine the type of incident, as well as a response, reporting, and recovery strategy. This will be covered in detail in an upcoming chapter.

Additional areas to consider are preventative measures against malicious attacks (e.g., DOS, theft, social engineering), patch and vulnerability management, configuration and change management (i.e., track all versions and changes to processes, hardware, software, etc.), and the fault tolerance of the components of any and all devices.

## 2.15   INFORMATION SECURITY GOVERNANCE AND RISK MANAGEMENT

Information security governance is the organizational structure to implement a successful information security program. To apply security governance (e.g., processes, policies, roles, compliance), one must first understand the organization's goals, mission, and objectives. Once these are understood, one can identify the assets of the organization and implement an effective risk management plan using tools to assess and mitigate threats to and vulnerabilities of the asset.

Risk assessment can be accomplished either quantitatively, qualitatively, or both. A quantitative example is being able to put a dollar amount on a risk. For example, if 1,000 records of patient data were exposed and it costs $30 to contact a patient, change his or her account number, and print a new health card, then the loss with this risk is $30,000 (Sims 2012). A qualitative risk identifies characteristics about an asset or activity (Gregory 2010):

- *Vulnerabilities:* An unintended weakness in a product
- *Threats:* An activity that would exploit the vulnerability
- *Threat probability:* The probability that the threat will occur (low, medium, high, or a numeric scale)
- *Countermeasures:* Tools to reduce the risk associated with the threat or vulnerability

Other considerations include managing personnel security and developing security training and awareness. Secure hiring practices such as performing reference checks, verifying education, and using employment agreements and policies between the employer and employee (e.g., nondisclosure) should be in place. In addition, once the person is hired, there should be security education, training, and awareness to mitigate risks. Training will be detailed in the "Preparing for an Incident" chapter.

In addition to software patches and fixes to protect against security vulnerabilities, sound judgment and caution are needed (Microsoft 2012). Here are the ten immutable laws of security according to Microsoft:

**Law #1:** *If a bad guy can persuade you to run his program on your computer, it's not your computer anymore.*

**Law #2:** *If a bad guy can alter the operating system on your computer, it's not your computer anymore.*

**Law #3:** *If a bad guy has unrestricted physical access to your computer, it's not your computer anymore.*

**Law #4:** *If you allow a bad guy to upload programs to your website, it's not your website anymore.*

**Law #5:** *Weak passwords trump strong security.*

**Law #6:** *A computer is only as secure as the administrator is trustworthy.*

**Law #7:** *Encrypted data are only as secure as the decryption key.*

**Law #8:** *An out-of-date virus scanner is only marginally better than no virus scanner at all.*

**Law #9:** *Absolute anonymity isn't practical, in real life or on the web.*
**Law #10:** *Technology is not a panacea.*

## REFERENCES

Conrad, E. 2011. *CISSP study guide.* Waltham, MA: Syngress.

Dey, M. 2011. Business continuity planning (BCP) methodology—Essential for every business. *IEEE GCC Conference and Exhibition*, February 19–22, pp. 229–232.

Dickson, B. 2016. The IoT ransomware threat is more serious than you think, IoT Security Foundation. https://www.iotsecurityfoundation.org/the-iot-ransomware-threat-is-more-serious-than-you-think/ (retrieved March 9, 2002).

Evans, D., Bond, P. and Bement, A. February 2004. Standards for security categorization of federal information and information systems. FIPS PUB 199.

Felten, E. July 24, 2008. What's the cyber in cyber-security? *Freedom to Tinker.* https://freedom-to-tinker.com/blog/felten/whats-cyber-cyber-security/ (retrieved August 12, 2012).

Grassi, P., et al. June 2017. Digital identity guidelines: Authentication and lifecycle management. NIST Special Publication 800-63B. https://nvlpubs.nist.gov/nistpubs/SpecialPublications/NIST.SP.800-63b.pdf.

Gregory, P. 2010. *CISSP guide to security essentials.* Boston: Course Technology, Cengage Learning.

Ikeda, S. March 25, 2020. IoT-Based DDoS Attacks are growing and making use of common vulnerabilities, *CPO Magazine.* https://www.cpomagazine.com/cyber-security/iot-based-ddos-attacks-are-growing-and-making-use-of-common-vulnerabilities/ (retrieved March 9, 2022).

Jaeger, T. 2008. *Operating systems security.* San Rafael, CA: Morgan & Claypool.

Khan, M. and Zulkernine, M. 2008. Quantifying security in secure software development phases. *Annual IEEE International Computer Software Applications Conference.*

King, R. July 6, 2009. Lessons from the Data Breach at Heartland, *Bloomberg Business Week.* http://www.businessweek.com/stories/2009-07-06/lessons-from-the-data-breach-at-heartlandbusinessweek-business-news-stock-market-and-financial-advice (retrieved August 9, 2012).

Locke, G. and Gallagher, P. 2009. Recommended security controls for federal information systems and organizations. NIST special publication 800-53.

Madou, M. January 11, 2021. SQL injection: The bug that seemingly can't be squashed, Helpnetsecurity.com. https://www.helpnetsecurity.com/2021/01/11/sql-injection-bug/ (retrieved March 9, 2022).

Martin, B., Brown, M., Paller, A. and Kirby, D. 2011. 2011 CWE/SANS top 25 most dangerous software errors. The MITRE Corporation.

Microsoft. 2012. 10 Immutable laws of security. http://technet.microsoft.com/library/cc722487.aspx (retrieved September 8, 2012).

NIST Joint Task Force. March 2020. Security and privacy controls for information systems and organizations. SP 800-53. https://nvlpubs.nist.gov/nistpubs/SpecialPublications/NIST.SP.800-53r5-draft.pdf (retrieved March 8, 2022).

Payne, J. 2010. Integrating application security into software development. *IT Professional* 12(2): 6–9.

Sims, S. 2012. Qualitative vs. quantitative risk assessment. SANS Institute. http://www.sans.edu/research/leadership-laboratory/article/risk-assessment (retrieved December 27, 2012).

Swanson, M. and Guttman, B. 1996. Generally accepted principles and practices for securing information technology systems. NIST special publication 800-14.

Talukder, A. K., Maurya, V. K., Santosh, B. J., Jangam, E., Muni, S. V., Jevitha, K. P., Saurabh, S., Pais, A. R. and Pais, A. 2009. Security-aware software development life cycle (SaSDLC)—Processes and tools. *IFIP International Conference on Wireless and Optical Communications Networks. WOCN '09*, pp. 1–5.

Theoharidou, M. and Gritzalis, D. 2007. Common body of knowledge for information security. *IEEE Security and Privacy* 5(2): 64–67.

*Title 44 United States Code Section 3542.* US Government Printing Office. http://www.gpo. gov/fdsys/pkg/USCODE-2009-title44/pdf/USCODE-2009-title44-chap35-subchapIII-sec3542.pdf (retrieved August 15, 2012).

Tracy, M., Jansen, W., Scarfone, K. and Butterfield, J. 2007. Guidelines on electronic mail security. NIST special publication 800-45.

Trend Micro. September 28, 2021. IoT and ransomware: A recipe for disruption, *Trend Micro*. https://www.trendmicro./vinfo/us/security/news/internet-of-things/iot-and-ransom ware-a-recipe-for-disruption (retrieved March 9, 2022).

Vijayan, J. 2010. Update: Heartland breach shows why compliance is not enough. *Computerworld*. http://www.computerworld.com/s/article/9143158/Update_Heartland_ breach_shows_why_compliance_is_not_enough (retrieved August 9, 2012).

Whitman, M. and Mattord, H. 2012. *Principles of information security*. Stamford, CT: Cengage Learning, Course Technology.

Young, S. January 26, 2022. Memorandum for the heads of executive departments and agencies, Executive Office of the President. https://www.whitehouse.gov/wp-content/ uploads/2022/01/M-22-09.pdf (retrieved March 9, 2022).

# 3 Strategy to Outpace the Adversary[1]

He who can handle the quickest rate of change survives.

—*John Boyd, Military Strategist*

## 3.1 INTRODUCTION

If the question "Are We Secure Yet?" is something you hear from your management team or they expect to see a checklist to show that all the right things were in place to secure the company assets, you may want to purchase this book for them. If management also expects metric massaging so they can tell a good story to company stakeholders about the secured company assets, instead of explaining the REAL cyber threat situation, you should give them several copies of this book to hand out to the stakeholders.

In the fast-paced arena of cyber-attacks, without keen strategic insight, the threat to data outpaces our ability to protect it. In this chapter we will share how to apply a time-tested air-combat strategic process and solid information security tactics to keep your data safe and your organization ahead of the cyber criminals. As you venture through this chapter, you will learn what will put your company in a better position to protect itself from the cyber adversaries faced daily. This is unlike the organizations in our field that pay more attention to metrics and dashboards to show "we are green" or "on the path to green" because that mentality implies that the real issues that are threatening our data and our systems are being ignored.

## 3.2 THE PROBLEM

If you have ever read the old tale of the Emperor's New Clothes, you will see it is analogous to the weak cyber security strategies employed at some organizations. The Emperor's New Clothes, a story by Hans Christian Anderson, is about an emperor that hired tailors to make him suits that are invisible to everyone that is incompetent. His followers pretended to see the clothes, although the emperor was naked. Even the tailors reinforced the concept that the clothes existed. This emphasizes the concepts where honesty is not valued in some organizations. Everyone goes along with the charade as they fear retribution if the obvious faults are pointed out (Neill et al. 2012).

The chief information security officer of an organization may have more interest in a compelling story to tell the board about "Yes we are Secure," instead of facing the real threat. They may believe that as long as we can present a secure face to the world, the adversary will go after our competition, after all, we only have to run faster than the other people that the bear is chasing. But what if there are more bears than there are people? We haven't reached that point—yet, however; the world is such a target-rich environment for hackers—so this phase will not last long if cyber security measures are not more strategic. These days, hackers are collaborating with each other, and toolkits using artificial intelligence are readily available, thus, creating a situation where the bears are getting faster and could outnumber the runners soon.

If you are looking for a strategic cyber security advantage, that will allow you to leverage the assets with a thorough understanding of your organization's risk appetite, then you may just learn that running from the bear is not your best course of action. Instead, executing specific tactics that come from a strategic view that will allow you to not just outrun the bear, but halt him in his pursuit of your organization's crown jewels will be much more effective. One must also prepare themselves for the possibility that those around you, or in your management will not have the courage to be a leader, to follow real strategy as they will feel threatened by your clarity and insight. They will do anything and everything in their power to show that the Emperor's New Clothes are amazing.

The goal of this chapter is to break through that existing pluralistic ignorance (*a social phenomenon where although an individual disagrees about something they go along with what they feel is the prevailing position of the group*) and introduce a cyber security strategy using the military decision-making model called the OODA Loop.

## 3.3  BOYD'S OODA LOOP OVERVIEW

John Boyd, a US Air Force pilot and military strategist, created a widely used decision-making model for high-state situations called the OODA Loop (Lewis 2019). Figure 3.1 shows the main components of the OODA Loop: *Observe, Orient, Decide, and Act.*

Boyd developed this concept in the cockpit of an F-86 Sabre jet fighter aircraft. During the Korean war, his team of fighter pilots shot down Soviet-made MiG-15 jets at a ratio of eight to one, despite the MiG-15 being a superior jet than the F-86.

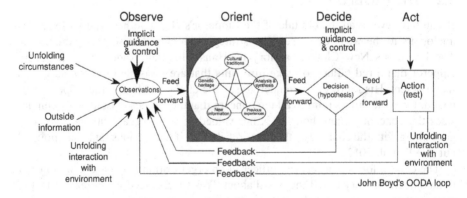

**FIGURE 3.1**   John Boyd's OODA Loop (https://commons.wikimedia.org/wiki/File:OODA. Boyd.svg).

How could that be? The difference was in how the cockpit was designed in the F-86 aircraft. The cockpit is described as having a high bubble canopy and a fully hydraulic flight control system. This enabled the US pilots to respond more quickly as they had better visibility and could shift and perform maneuvers more quickly with this design (Grazier 2018). The OODA Loop concept is based off of the concept of having a clearer focus.

Boyd defined strategy as:

> A mental tapestry of changing intentions for harmonizing and focusing our efforts as a basis for realizing some aim or purpose in an unfolding and often unforeseen world of many bewildering events and many contending interests.

Boyd further defined the purpose of strategy to be able to improve our ability to be adaptable to unexpected circumstances. The following sections explain the four OODA Loop steps. The chapter ends with a real-world application of the OODA Loop applied to the Ukraine-Russia cyberwar conflict.

### 3.3.1 Observe in Depth

You have to be able to measure and observe consistently in order to identify threats. This can be the most daunting part of the OODA Loop. As an example, at the Commonwealth of Pennsylvania, a Security Information and Event Management report/observational tool was implemented. The original implementation only collected security events. As the Security Information and Event Management marketplace matured, Pennsylvania realized that they needed a correlation of these events—thus data analysis to enable rich reporting and alerting was incorporated into the tool.

Due to technology advances, we have so many pieces of information that we are trying to observe, that the information garnered can become meaningless. Thus, a deeper understanding of the organization's core business is required. To get that type of understanding, one needs to think about the following three items: (1) what assets your company has; (2) what are web facing assets; and (3) which of those *need to be* facing the web (in reality).

This involves reading between the lines. In fact, that is where cybercrime resides— *between the lines*. In other words, you must think like a bad actor. The bottom line is understanding all of the potential threats to your business. For example, that means thinking about what is web facing and how can it be protected? Such as which ports are open, what data is critical, where does the critical data reside, and any other elements in your environment that cause system and data vulnerability.

In addition, third-party partners that are part of your cyber ecosystem need to be considered. Third-party services run the gamut from data services to matching/data enhancement to improve operations. Therefore, the partners that have access to the systems that process the company data needs to be addressed.

Observations have very little value if they are simply a point in time, such as yearly or longer. Regulations (such as New York State DFS 500) are requiring continuous or near real-time monitoring (observations). In other words, to be strategic you have to expand your thought process—not just what's happening on your network—but consider where your data lives.

Today, the crown jewels of an organization are all electronic, that includes intellectual property. Cyber security is so much more than identity and tangible theft (such as a mechanical device being stolen by a former employee where a forensic investigation ensues to determine if an employee stole an AutoCAD drawing for example).

This brings us to the $10 billion question—How/what visibility can be gained from an organization's down chain parties? Today's best practice to gain that visibility is via a contract. The state of the industry surrounding third party, data sharing, and fourth-party engagements may be to require a third party to conduct (and share) fourth and down chain assessments. One suggestion is to put some type of device or control in place in real time. However, this poses liability and other risks (including privacy). It is clear that in today's technology landscape, the process of observation needs to be automated and near real time.

### 3.3.2 ORIENT IN DEPTH

Orientation is the heart of the OODA Loop process. During orientation we take our observations and apply previous knowledge, existing correlations, and even intuition. Essentially it is trying to make sense of what we are seeing in order to drive a decision. In the world of cyber security, there is far too much data for human orientation. Thus, machine learning is crucial in assisting seeing the trees in the forest. When you are researching tools to assist in this process, be wary of vendors who claim they produce actionable intelligence. While they may provide intelligence and perhaps assistance in assimilating it—taking action without orienting it to your company's risk appetite, assets, legal and regulatory requirements, and other information that you may have—can result in errors and churn. Always orient the intelligence.

### 3.3.3 DECIDE IN DEPTH

Decide in depth is what you do after you have consumed (observed) and digested (orient) a significant amount of data. Let's assume that as part of your overall cyber strategy in continuously monitoring your supply chain, you have reviewed a number of vendors who provide what they market as "continuous monitoring services." You simply pass them the information of a particular supplier and they do the rest. They grab public financial records, regulatory findings, web presence vulnerability reports, custom analysis, and hands on review of the data and they pass you a score, number, or another piece of information that tells you that this vendor has become problematic and that their risk to your cyber security has suddenly changed in a negative fashion—in other words, a trigger that will change your organization's posture with this vendor has occurred. What is next?

In the typical scenario, you may decide that something needs to be done to protect your corporate data and its brand. Or you may have gotten funding for this tactical measure (the contracted manufacturer's vendor product) by using the vendor's marketing material about return on investment, or securing your supply chain, or preventing cyber events before they happen. Current regulations (as of this writing) such as the NYS DFS 500 regulation require organizations to perform some type of continuous monitoring of third parties and this tactical solution fits the bill perfectly, checks the box, and according to the vendor reduces your risk and makes you more secure. Therefore, the vendor's marketing campaign may actually be right on target, and the services they provide will give you actionable intelligence that you can use to keep the castle safe. The real question is what your strategy tells you to do with the intelligence (observation) and recommended action (orientation performed by the vendor on your behalf). A strategy like OODA Loop provides a governance structure so individuals can't get side-tracked when taking security measures.

You now have the power to make a decision—remember to go back to the other parts of the OODA Loop to help drive your decision making. This is where play-books and frameworks have high value for your process of decision making and act-ing. In fact, you might make the decision to stop a contract, or you may take another path depending on what's in play in the particular situation.

The success in your OODA Loop resides in having strong (thoughtful, structural) governance to drive your decision making. It can be burdensome if there is too much governance. For example, the first stages of the war during the Blitz bypassing the Maginot Line that the Germans essentially defeated the French in 2 weeks. The French were using their WWI OODA Loop—they gathered info and then sent it up the chain of command for a decision to be made—this slows down the decision process and removed the observation from the currency (the situation at hand). Your observation may have changed by the time you get back the higher command—the difference for the Germans in WWII gave their field command the power to make decisions and proceed immediately based on their immediate observations.

The playbook should be devised with as many options as possible (as many pre-determined situations) to allow for performing an action (*we will refer later to a case study on phishing in Chapter 10—it is not as bad as it used to be, numbers are down but, people are still getting caught*). The fallacy of awareness training—a specific step needs to be available to derail phishing in that case.

The team needs the power to make a decision based on any observation—the more data you have, the more decisions you have to make—if the governance struc-ture is not clear and agile, the decision to act will be bogged down and the bad event cannot be stopped, instead of acting in agile manner and actually stopping the bad event.

### 3.3.4 ACT IN DEPTH

Once you have that good governance and you are able to act—the actions you take (this is part of the playbook and governance concept) will be clearly defined. The experts will not always be the ones in the command center (e.g., the ones who have done the orientation phase), so those on the ground have to understand what actions to take, not just when to decide to act.

Before the problem ever arises—in a mature program, all the elements to answer these questions are in place:

- Who do you contact?
- How do you escalate?

- Is this automated?
- If there are outliers, how are they to be dealt with?
- What communications need to take place?

Feedback involves acting in depth. This resets the loop once you've done your action. The next step is observing the results of that action—the O in the original OODA. This also speaks to governance.

## 3.4  OODA LOOP APPLIED: THE UKRAINE-RUSSIA CYBERWAR

We are seeing cyberwar as we have never seen before. In this section we will apply the OODA Loop to the cyberwar caused by the 2022 Ukrainian and Russia conflict. In this example, we will focus on how to protect an organization when their ecosystem includes multiple vendors—that could possibly be a part of a global cyberwar.

In early 2022, state-sponsored hacker groups publicly disclosed which flag they're representing in a cyber battle. The notorious Conti Ransomware Gang (https://blackkite.com/the-conti-playbook-leak-your-questions-answered/) declared their support for Russia and intentions to execute cyber-attacks within their full power against countries supporting the Ukraine. Additional volunteer underground hacker groups, including Anonymous, asserted their counter shortly after, supporting Ukraine and their stance against Russian institutions (both assertions are shown in Figure 3.2).

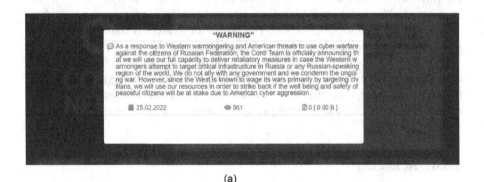

(a)

**Anonymous** @YourAnonOne                                          •••
The Anonymous collective is officially in cyber war against the Russian government. #Anonymous #Ukraine

(b)

**FIGURE 3.2**   Retrieved from the dark web, February 2022.

These bold escalations increased concerns significantly for many companies worldwide about the safety of their digital supply chain. In this example, we will use Black Kite's platform (https://blackkite.com/platform/) within the OODA Loop process to demonstrate visibility to an organization's customers around the overall cyber risk of any vendor, including ransomware susceptibility and distributed denial of service (DDoS) attack resiliency in relation to Ukrainian and Russian-based assets.

Not only are these two countries executing state-sponsored attacks on each other, but as mentioned, bad actor groups and ransomware gangs have taken sides in the war. From a business perspective, Black Kite's ecosystems contain many vendors that provide critical services. The reality is that some of those vendors may have IT assets or offices located in the area of conflict. Which begs the question, how do we understand what the potential impact would be if one of those suppliers became a victim to a cyber-attack?

This is where the OODA Loop presented in this chapter comes into play. In this example, the goal would be to understand where an organization's vendors' assets are actually located. Historically, the way that was done by collecting observations by communicating with all critical vendors asking them to self-identify if they have assets or locations in a specific country. This process is very time-consuming as it takes an extensive amount of time to collect and correlate these observations before they could even be subject to analysis in the orientation phase of the OODA Loop.

In this case, what we will show is how a cyber security company called Black Kite has automated this process and can be used within the OODA Loop practice. The Black Kite tool uses a methodology that takes observations and automatically correlates the connections between those assets and their geo locations making it easier and faster to make actionable decisions. Black Kite, as part of its business model, collects information about technical controls that are exposed on the Internet for every company on the planet. The tool's algorithms perform analysis resulting in correlations to assist an organization in making informed decisions.

### 3.4.1 OBSERVE

In this first phase of the OODA practice, the collection of all the controlled data is observed. With multiple attack vectors and malware used in the Russian–Ukrainian cyber-attacks, such as HermeticWiper and Cyclops Blink (replacement of VPNFilter malware), there is no single way of protecting your vendor ecosystem against escalating cyber threats.

Some countries publicly stated their support for either side of the crisis, likely ranking themselves higher on threat actors' hit lists. Vendors from these countries might be more subject to cyber-attacks, especially vendors residing and owning IT assets in Ukraine and/or Russia. Black Kite can determine vendor risk by identifying the location of IT assets of the companies and also their vendors (fourth parties). With smart tags, Black Kite alerts customers to the impact of cyber incidents related to the Russia-Ukraine war on their vendor ecosystems (Figure 3.3). The vendor risk dashboard displays a significant amount of information about each vendor, information like cyber hygiene scores, ransomware susceptibility, probable financial impact from a cyber incident, compliance levels, and more, and this use case focuses on a

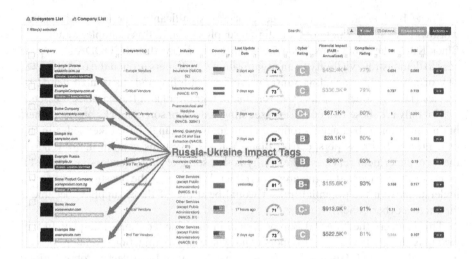

**FIGURE 3.3**   Vendor risk dashboard.

rapidly emerging threat. The Impact tags specific to this threat allow an analyst to triage the most impactful vendors.

### 3.4.2  ORIENT

The orientation phase is using the algorithms to analyze for a specific result, which is understanding which companies have IT assets or physical locations in Russia or Ukraine. That data is then flagged and presented to the user so it can be easily understood and actioned.

As shown in Figure 3.3, companies have either one or multiple of these tags and in some cases zero. As the orientation continues, one would look at those companies and go back to the observation phase to gather additional information about the relationships you have with those companies, as well as the data that is exposed. From the first round of observations (Black Kite's database), the process has taken many companies out of scope for further analysis simply because they don't fit the goal of your analysis.

#### 3.4.2.1  Understand the Cyber Security Posture of Your Third and Fourth Parties

Threat actors might advance their way upwards in the supply chain by initially attacking a fourth-party vendor (vendor of a vendor). Even if a vendor is not directly related to targeted countries, your company might still be impacted due to a targeted fourth party.

In addition to smart tags, Black Kite follows and applies commonly used frameworks developed by the MITRE Corporation to score software weaknesses consistently and transparently, converting highly technical terms into simple letter grades (Figure 3.4). The vendors with lower grades can appear as easy opportunities to threat actors, and in turn, quickly be targeted. Organizations can also share technical

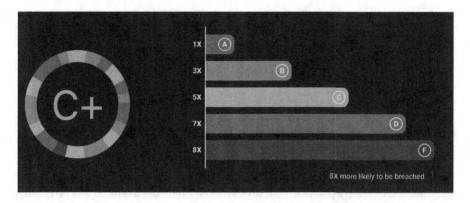

**FIGURE 3.4**  Cyber hygiene grading.

details with their vendors with one click, alerting business partners to vulnerabilities and allowing them to improve accordingly.

### 3.4.2.2  Understand the Ransomware Susceptibility of Your Vendors

The Conti ransomware group is not alone in attacking organizations they perceive as "enemies." Methods such as weaponizing data-wiper malware to extort companies, or simply hinder their operations, emphasize the need to monitor vendor ransomware risk and the anomalies in their ecosystems.

Indicator of Compromise lists are available in open sources. Playbooks of infamous gangs analyzed by research groups can also help understand their tactics, techniques and procedures (TTPs) For instance, Black Kite research recently published three analyses: (1) Conti's methods (https://blackkite.com/-conti-catches-fire-as-ransomware-business-model-matures/), (2) playbook (https://blackkite.com/the-conti-playbook-leak-your-questions-answered/), and (3) victims (https://blackkite.com/contis-ransomware-spree-last-week-half-of-the-victims-are-manufacturers/). The Cybersecurity and Infrastructure Security Agency also published an alert about the destructive malware targeting organizations in Ukraine, including technical details (https://www.cisa.gov/uscert/ncas/alerts/aa22-057a).

### 3.4.3  DECIDE

The next step is to dig deeper into the control data, combining that with the relationship data and deciding if there would be an impact to you in the event of a cyber-attack on that company. This is the decision phase of the OODA Loop.

Black Kite's Ransomware Susceptibility Index (RSI™) (https://blackkite.com/-ransomware-susceptibility-index/), shown in Figure 3.5, follows a process of inspecting, transforming, and modeling collected from various OSINT sources (Internet-wide scanners, hacker forums, the deep/dark web, and more). This data was also shown in the dashboard (Figure 3.3). Using the data and machine learning, the correlation between control items is identified to provide approximations.

**FIGURE 3.5**   Black Kite's Ransomware Susceptibility Index (RSI™).

With the RSI™, organizations can understand which vendors are most prone to ransomware and develop a practical course of action for remediation by cross-correlating findings with Black Kite's Cyber Risk Assessment.

### 3.4.3.1   Monitor the DDoS Resiliency of Your Vendors

Threat actors exploit DDoS attacks to block services and interrupt business. This method has become a primary weapon used by state-sponsored hackers to paralyze the target systems.

Russian hackers used this DDoS cyber weapon to attack Ukrainian state banks and public institutions on both February 15 and 23, 2022. The attacks resulted in service disruptions for hours, creating a ripple effect throughout the government institutions as a whole.

It is possible adversaries might use the same weapon for companies to target your vendor ecosystem. Therefore, monitoring DDoS resiliency is imperative. Begin asking vendors to block suspicious IP addresses, such as IP addresses in the blacklists due to botnet infection and IP addresses originating from Russia and other wary countries. Threat actors are suspected of using virtual private network (VPN) and the onion router (ToR) services to hide their tracks, so blocking IP ranges of Free VPN and ToR services is meaningful.

One of the 20 Black Kite Cyber Risk Assessment Categories (https://blackkite.com/technical-grade/) is DDoS Resiliency, which results from 15 different potential DDoS checks and detects any potential DDoS amplification endpoints. The data is collected from non-intrusive scanners and other Internet-wide scanners.

Organizations can filter vendors with respect to their DDoS ratings and inform vendors rating lower about the misconfiguration.

### 3.4.4   Act

After an actionable decision is made, you would then follow your best practices, if you have, them, on your incident response with your vendors. Those actions could include many different things such as arranging for a new duplicative service if a critical service was lost because of a cyber-attack in Ukraine. It could also be an action where you reach out to the vendor to further gather information and understand what they are doing to ensure that they do not become a victim. It's the combination of all the things that we've just discussed in this use case that clearly demonstrate the value of the OODA Loop.

### 3.4.5 TAKE AWAY POINTS

There are five takeaway points or lessons learned from this example:

1. Monitor your third and fourth parties' cyber security posture
2. Identify possible targets in your vendor ecosystem (IT asset locations might help)
3. Confirm vendors have the Indicator of Compromise list of possible malware in use by state-sponsored hackers
4. Monitor the ransomware susceptibility of your vendors. Ensure they patch recent critical vulnerabilities, use up-to-date services, do not have critical open ports such as remote desktop protocol (RDP) or server message block (SMB), and correct email configuration to mitigate phishing attacks
5. Monitor the DDoS resiliency of your vendors and make sure they block suspicious IP addresses

## 3.5 CONCLUSIONS AND RECOMMENDATIONS

The word loop in the title indicates that the OODA process is not a linear progression, it can be iterative. For example, during orientation you may find you do not have enough pertinent observations. The appropriate decision may be to go back to your observations and collect more data. Also, you may have developed implicit guidance. Implicit guidance is "an unconscious preplanned physical response to a known threat stimulus, which is often referred to by psychologists as a 'learned automatic response'" (Stephens 2013). The response is based on previous experience which allows one to progress through the OODA Loop efficiently by going directly to the action phase.

As mentioned earlier, "Are we secure yet?" is the question that shows a lack of understanding of the true issues at hand. Cyber security is essentially the battle between the professionals working diligently to protect our digital assets, and the world of cyber criminals who are constantly working to break through our defenses and monetize the breach. The professionals have processes to follow, budgets to acquire, and programs to build and run. However, the bad actors simply adapt very quickly and learn from their failures to improve their criminal activity.

Why are we hearing the same solutions over and over and yet the number of breaches is increasing, and cyber risk is more prevalent? The threat to our very identity and safety is worse. Why is this? As a chief information security officer, chief risk officer, or a board member, your process for becoming secure takes a long time. It could take 6 months to 1 year even if you use an outside provider to help secure your organization. This is the tactical limitation of developing processes to block a cyber threat. The bad actors don't have all the limitations that the professionals have. The cyber criminals are looking for the next easy target, just like the traditional burglar who will skip the house with the barking dog, with the security alarm sign, outdoor cameras, and extra home lighting, because they are looking for the opportunity—the easy target—the house with no dog or system. Today, if one is attempting to protect their assets, the criminals will back off and look for another victim. This is what makes third parties, such as the supply chain, so attractive. Cyber security is not typically on a third parties' agenda. We need to get our strategic and operating loops more efficient and effective than they are now.

## REFERENCES

Grazier, D. September 13, 2018. Why the OODA loop is forever. https://taskandpurpose.com/thelongmarch/military-ooda-loop-forever/ (retrieved December 14, 2021).

Lewis, S. June 2019. OODA loop. https://searchcio.techtarget.com/definition/OODA-loop (retrieved December 14, 2021).

Neill, C., Laplante, P. and DeFranco, J. 2012. *Antipatterns managing software organizations and people.* CRC Press.

Stephens, D. September 19, 2013. Understanding the OODA loop. *Police Magazine.* https://www.policemag.com/341024/understanding-the-ooda-loop (retrieved February 2, 2022).

## NOTE

1 This chapter was contributed by Bob Maley and Dr. Ferhat Dikbiyik.

# 4 Preparing for an Incident[1]

Before anything else, preparation is the key to success.

*—Alexander Graham Bell*

## 4.1 INTRODUCTION

It has been said, "If you don't prepare the garden in the spring it won't be ready to harvest in July." Preparing does not just include turning the ground over and planting the seeds; it also includes putting up a fence and using deterrents for those pesky deer and rabbits trying to eat everything. If we apply that sentiment to security incidents, the information security professional obviously needs much more than a firewall to deter those pesky hackers from our critical data and systems as was outlined in the previous chapter. In addition, the digital forensics professional also needs the security deterrents set up in such a way that they will have the appropriate means to help catch and convict the intruder if he or she does get unauthorized access. The digital forensic process also needs to be able to gather information to determine and secure the vulnerability that was exploited. Thus, in this chapter, the topics and safeguards that needed to be implemented preincident to facilitate securing the assets as well as provide data for an effective investigation postincident will be discussed.

It can be an overwhelming task for any business—well aware of the exponentially growing cyber threats to its critical assets—to maintain a comprehensive security posture. A comprehensive security posture will include features that will not only mitigate incidents but will also make both incident response (IR) and the digital forensic investigation much more effective because all bets are that the incident will occur eventually and will need to be investigated. In other words, just as in firing a weapon, "aim, fire" is not enough, the weapon should be "ready" too. In an organization in which resources are tight, preincident preparation and the development of the IR process may take a back seat to responsibility of securing the system. The point is that a successful preincident preparation process should have its own focus by *both* the information security *and* digital forensics professionals to reduce the cost and increase the success of an investigation. In particular, this focus will help ensure that policies are followed and that data collected can be used as evidence in a court of law if necessary.

I chose to present this topic using a framework that can help communicate how a company can prepare to protect its infrastructure and critical data from cyber threats as well as be in a position to perform an effective digital forensic investigation if and when an incident occurs. The proposed framework is notably derived from the Zachman enterprise architecture (Zachman 1987).

Before we dive into the framework it is essential to understand the foundations of digital forensics. The National Institute of Standards and Technology (NIST) provided a report (NISTIR 8354-DRAFT) presenting the foundations of digital forensics. This is an important topic as there are many ways to search for digital information—but those "ways" need to be trusted (Lyle et al. 2022). Here are 12 takeaways from the NIST report:

1. In routine operations computers store much more data than what is presented to the user. Examples include storing time and location data on photos, extra copies of data, and data about system activities. Forensic tools and techniques can reveal this data to provide a window into activities that have taken place on a computer or other digital device.
2. Digital forensics is dependent on an understanding of computers and how they work. Any activity that is performed by a computer can potentially be a target for a forensics tool or technique.
3. Computer technology evolves rapidly but sporadically. Some attributes of computers last for decades and some only for a few weeks.
4. The forensic examiner needs to be aware of key changes in computing technology relevant to the examination being performed. Frequent changes in digital technology introduce the possibility for incomplete analysis or for misunderstanding of the meaning of artifacts.
5. Not every digital forensic technique undergoes a peer review, formal testing, or error rate analysis. In general, the digital forensics community performs an informal review by providing feedback about the usefulness of techniques. This general acceptance process allows for techniques to be quickly evaluated and revised.
6. When using techniques to recover deleted or hidden artifacts the examiner must determine the relevance of the recovered information as it may be incomplete or improperly merged with irrelevant information.
7. Searching tools have limitations based on the multiple ways that computers store information. Limitations include the type of files, types of encoding, and many other parameters. In general search tools are very effective at finding information, but there is a possibility that data will be missed because a tool does not have the capability to find it.
8. If someone has taken steps to change information in digital evidence to mislead an examiner, it may be difficult to detect the changes. Depending on the sophistication of the manipulation, identification of the changes relies on the skill of the examiner.
9. Digital processes tend to have systematic errors rather than random errors. Therefore, an error mitigation analysis provides more information and is the correct way to manage uncertainty. Asking for an error rate is only useful where there are random errors.
10. When error rates are provided, it is important for the user to understand the context of the numbers. Errors in computer science techniques tend to be so small as to be negligible. For some forensic techniques, the error rates may vary significantly based on attributes of the technology and usage patterns.
11. It is not feasible to test all combinations of tools and digital evidence sources.
12. Extensive tool testing of over 250 widely used digital forensic tools showed that most tools can perform their intended functions with only minor anomalies.

### 4.1.1 THE ZACHMAN FRAMEWORK

The widely used Zachman framework (Figure 4.1), developed by John Zachman in 1987, provides a way to rationalize architectural concepts and facilitate communication among the designers of complex information systems. The Zachman framework can be adapted to model a variety of complex systems.

In the figure, the six dimensions shown as the rows in the framework represent different stakeholder perspectives of a complex system. Essentially, the framework lays out the architectural model including each stakeholder in a system, thus creating a complete view of that system. Zachman's main goal was to emphasize the fact that systems design is not just about the system itself; instead, it is an enterprise issue.

The columns in the framework are a metamodel that answers the questions, what, how, when, who, where, and why to describe the enterprise.

### 4.1.2 ADAPTATION OF THE ZACHMAN FRAMEWORK TO INCIDENT RESPONSE PREPARATION

The only reported application of the Zachman framework to digital forensics is Leong's (2006) FORZA model. In this case, Leong used Zachman to define eight different roles and responsibilities in a digital forensic investigation via a set of interrogative questions that can be utilized during an investigation. While Leong's model provides a rigorous way to approach postincident data collection, the FORZA model does not address the issue of preparation, which will have an effect on the success of the investigation. Mandia, Prosise, and Pepe (2003), a highly cited resource in IR, recommended six areas to be addressed in preincident preparation: identifying risk, preparing hosts, preparing networks, establishing policy/procedure, creating a response toolkit, and creating a team to handle incidents. With the addition of a new dimension, *training,* coupled with several special publications of the NIST and a few other resources, the Zachmam framework is modeled for the digital forensics preparation process (DeFranco and Laplante 2011). This new framework provides a model to analyze the vulnerabilities critically, gives suggestions for security and education, and presents a plan for the overall protection of an enterprise's resources, data, and information.

In the case of preincident preparation, these seven abstraction layers can be derived by analyzing the package diagram shown in Figure 4.2.

Table 4.1 illustrates the areas that were adapted by applying the Zachman framework as well as the factors that are of interest to prepare effectively for an incident.

The following sections will highlight the activities described in Table 4.1 in further detail. Section headings correspond to row headings in the table, and the italicized questions are fully realized versions of the factor list in the second column of the table.

## 4.2 RISK IDENTIFICATION

*Why do we need to evaluate security threats continuously?*

The enterprise must contend with malware, vulnerabilities in their applications, user mobility, spam, even its employees—the list could go on and on.

## ENTERPRISE ARCHITECTURE - A FRAMEWORK ™

| | DATA *What* | FUNCTION *How* | NETWORK *Where* | PEOPLE *Who* | TIME *When* | MOTIVATION *Why* | |
|---|---|---|---|---|---|---|---|
| **SCOPE (CONTEXTUAL)** *Planner* | List of Things Important to the Business | List of Processes the Business Performs | List of Locations in which the Business Operates | List of Organizations important to the Business | List of Events Significant to the Business | List of Business Goals/Strat | **SCOPE (CONTEXTUAL)** *Planner* |
| | ENTITY = Class of Business Thing | Function = Class of Business Process | Node = Major Business Location | People = Major Organizations | Time = Major Business Event | Ends/Means = Major Bus. Goal/ Critical Success Factor | |
| **ENTERPRISE MODEL (CONCEPTUAL)** *Owner* | e.g. Semantic Model | e.g. Business Process Model | e.g. Logistics Network | e.g. Work Flow Model | e.g. Master Schedule | e.g. Business Plan | **ENTERPRISE MODEL (CONCEPTUAL)** *Owner* |
| | Ent = Business Entity Reln = Business Relationship | Proc = Business Process I/O = Business Resources | Node = Business Location Link = Business Linkage | People = Organization Unit Work = Work Product | Time = Business Event Cycle = Business Cycle | End = Business Objective Means = Business Strategy | |
| **SYSTEM MODEL (LOGICAL)** *Designer* | e.g. Logical Data Model | e.g. "Application Architecture" | e.g. "Distributed System Architecture" | e.g. Human Interface Architecture | e.g. Processing Structure | e.g. Business Rule Model | **SYSTEM MODEL (LOGICAL)** *Designer* |
| | Ent = Data Entity Reln = Data Relationship | Proc = Application Function I/O = User Views | Node = I/S Function (Processor, Storage, etc) Link = Line Characteristics | People = Role Work = Deliverable | Time = System Event Cycle = Processing Cycle | End = Structural Assertion Means = Action Assertion | |
| **TECHNOLOGY MODEL (PHYSICAL)** *Builder* | e.g. Physical Data Model | e.g. "System Design" | e.g. "System Architecture" | e.g. Presentation Architecture | e.g. Control Structure | e.g. Role Design | **TECHNOLOGY CONSTRAINED MODEL (PHYSICAL)** *Builder* |
| | Ent = Segment/Table/etc. Reln = Pointer/Key/etc. | Proc = Computer Function I/O = Screen/Device Formats | Node = Hardware/System Software Link = Line Specifications | People = User Work = Screen Format | Time = Execute Cycle = Component Cycle | End = Condition Means = Action | |
| **DETAILED REPRESENTATIONS (OUT-OF-CONTEXT)** *Sub-Contractor* | e.g. Data Definition | e.g. "Program" | e.g. "Network Architecture" | e.g. Security Architecture | e.g. Timing Definition | e.g. Rule Specification | **DETAILED REPRESENTATIONS (OUT-OF CONTEXT)** *Sub-Contractor* |
| | Ent = Field Reln = Address | Proc = Language Stmt I/O = Control Block | Node = Addresses Link = Protocols | People = Identity Work = Job | Time = Interrupt Cycle = Machine Cycle | End = Sub-condition Means = Step | |
| **FUNCTIONING ENTERPRISE** | e.g. DATA | e.g. FUNCTION | e.g. NETWORK | e.g. ORGANIZATION | e.g. SCHEDULE | e.g. STRATEGY | **FUNCTIONING ENTERPRISE** |

Zachman Institute for Framework Advancement - (810) 231-0531

**FIGURE 4.1**   The Zachman framework.

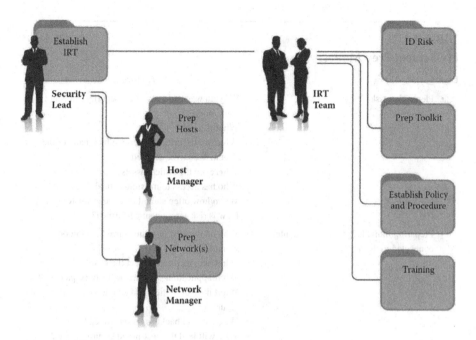

**FIGURE 4.2** Package diagram of the preincident prep framework.

Due to the ever-changing nature of the threat space, it is impossible to determine every vulnerability in a network; however, by determining and addressing the known vulnerabilities, the enterprise can be prepared both offensively and defensively.

An effective approach to improving an organization's security posture and preventing incidents is to conduct periodic risk assessments of systems and applications (Grance, Kent, and Kim 2004). Assessing risk is clearly the first step in improving a company's security posture.

*What is a risk?*

A risk is the probability that a threat will take advantage of a system's vulnerability.

*What are the critical assets (i.e., which parts of the system need to be secured)? Where are the critical assets located? Who has access to mission-critical information/data?*

Critical assets pertain to confidential customer or company data, critical plans, private individual data, or even the corporate reputation. Anything that, if stolen or compromised, would be harmful to the company's future is considered a risk. Although large networks can be vulnerable to hackers, defenders also need to worry about the malware located everywhere on the Internet that their users are perusing as well as external threats getting to the critical assets. Thus, the threats faced by most organizations now have an additional focus on internal users. Those users are a big risk due to their privileged access to confidential information as well as their access to critical applications.

## TABLE 4.1
## Preincident Prep Model

| Layers | Factors |
|---|---|
| • Identifying risk | • **Why** do we need to evaluate security threats continuously?<br>• **What** is a risk?<br>• **What** are the critical assets? Which parts of the system need to be secured?<br>• **Where** are the critical assets?<br>• **Who** has access to mission-critical data?<br>• **When**/how often should risks be evaluated?<br>• **How** is risk assessment performed? |
| • Preparing individual hosts (a computer connected to the network) | • **Why** do the host computers need continuous monitoring?<br>• **What** activities can protect the host?<br>• **What** application software needs to be patched?<br>• **What** data are critical and should be backed up and secured?<br>• **When** should backups be scheduled?<br>• **Who** will lead the host-based security effort?<br>• **Where** are the host computers are located?<br>• **Who** needs to be educated on host-based security?<br>• **Who** has access to the hosts?<br>• **How** can data be protected while using cloud services? |
| • Preparing the network | • **Why** does the network need continuous monitoring?<br>• **Who** will manage this effort?<br>• **What** activities can protect the network?<br>• **When** did the events occur? Network synchronization.<br>• **Where** did the incident occur?<br>• **How** to prepare for an advanced persistent threat (APT) |
| • Establishing appropriate policies and procedures | • **What** is an acceptable use policy (AUP)?<br>• **Why** do we need an AUP?<br>• **Who** is affected by the AUP?<br>• **How** do we determine what the AUP should address?<br>• **When** and how often should the AUP be updated?<br>• **Where** should the AUP document be kept? |
| • Creating/preparing response tools kit | • **Why** is a response toolkit necessary?<br>• **When** is the response toolkit used?<br>• **Where** is the toolkit used?<br>• **How** is the toolkit assembled?<br>• **What** is the necessary hardware, software, and documentation needed to respond to incidents? |

*(Continued)*

**TABLE 4.1 (*Continued*)**
**Preincident Prep Model**

| Layers | Factors |
|---|---|
| • Establishing an incident response team | • **Why** do we need to establish a response team? |
| | • **Who** should be on the computer incident response team (CIRT)? |
| | • **How** to establish a CIRT? |
| | • **When** should stakeholders be contacted? |
| | • **What** factors need to be considered in creating a CIRT? |
| | • **Where** is the incident evidence stored? |
| • Training | • **Why** is it important to educate the users? |
| | • **Who** will do the training? |
| | • **Where** should the training occur? |
| | • **When** should the training occur? |
| | • **How** should the users be trained? |
| | • **What** should the users be trained on? |

Creating and documenting a network topology can also assist in the risk assessment effort. The topology will show where the critical assets are located and how they are connected. This view can highlight vulnerabilities that need to be addressed.

*When/how often should risks be evaluated?*

Risks evolve, networks change, and people leave—all reasons why the risks need to be evaluated as often as possible. Vulnerability risk evaluation should be run weekly for optimum security and monthly—at least—for best practice (NTT 2009).

*How is a risk assessment performed?*

NIST suggests the following risk assessment methodology (Stoneburner, Goguen, and Feringa 2002):

1. *System characterization:* Understand the system topology and gather information to determine the system's environment and boundary.

2. *Threat identification:* Determine a list of threat sources that may exploit any of the system vulnerabilities or an action that may unintentionally initiate exploitation of a vulnerability. A threat source can be a hacker, a criminal, or even a poorly trained or unhappy employee.

3. *Vulnerability identification:* Determine the weaknesses that could be exploited. There are many known vulnerabilities that can be found through vulnerability databases,[2] vendors, vulnerability scanning tools, and penetration testing. Other vulnerabilities can be found by assessing the security requirements of the system itself.

4. *Control analysis:* Determine controls to minimize the probability of a security incident. Many of the access control techniques, such as authentication and encryption discussed in Chapter 3, are controls that may be utilized to minimize incidents.

5. *Likelihood determination:* To determine the likelihood of a threat source exercising the vulnerability, consider the motivation and capability of the threat source, the nature of the vulnerability, and the access controls in place. The likelihood can be rated as *high, medium,* or *low.*

6. *Impact analysis:* Determine the magnitude of the impact. An impact analysis as was described in Chapter 3 can be implemented to determine the effect of the incident on the integrity, availability, and confidentiality of the system and data. The impact can be rated as *high, medium,* or *low.*

7. *Risk determinations:* Determine the level of risk that the threat source will take advantage of the vulnerability by evaluating the likelihood of an attempt, the magnitude of the impact, and the effectiveness of the security controls.

8. *Control recommendations:* Determine effective controls and alternate solutions to reduce risk probability. These determinations should consider the effectiveness of the options, legislation/regulation, organization policy, operational impact, safety, and reliability.

9. *Results documentation:* Assemble an official report describing the analysis.

The Ponemon Institute surveyed 56 US organizations where the results showed that the cost of cybercrime averaged $8.9 million. This is a 6% increase over the 2011 study. Those companies experienced an average of 1.8 successful attacks per week—a 42% increase from 2011. The most common attacks are denial of service, malicious insiders (permanent and temporary employees, contractors, and business partners), and web-based attacks. Figure 4.3 depicts the comparison of the US cybercrime cost average against those of four other countries.

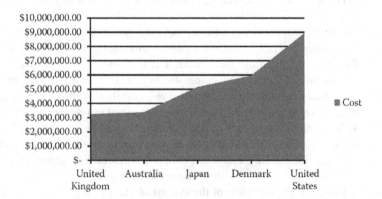

**FIGURE 4.3** Cybercrime cost comparison of five countries. (Adapted from Ponemon Institute, 2012, http://www.ponemon.org/local/upload/file/2012_US_Cost_of_Cyber_Crime_Study_FINAL6%20.pdf.)

## 4.3   HOST PREPARATION

*Why do the host computers need continuous monitoring?*

With the growth of mobile computing, there are many devices connected to the network. Devices that are connected to corporate networks now have access to sensitive data. In addition, they all connect to the Internet, cloud services, and social networking sites that are all significant contributors to security incidents.

*What activities can protect the host?*

NIST recommends several important practices for securing a host, such as limiting user privileges, evaluating default settings and passwords, displaying warning banners for unauthorized use, and enabling logging (useful for incident investigations) of significant security-related events (Grance et al. 2004).

Believe it or not, user password choice continues to be a problem. For example, Yahoo Voices, an open-publishing platform, had 453,491 plaintext passwords and e-mail addresses stolen (Brading 2012). The hacking was due to a structured query language (SQL) injection. What was discovered is a good reminder of why passwords need to be evaluated. These stolen passwords were FAR from secure; for example, 1,373 people used "password" as their password and 1,666 used "123456" (Cluley 2012). This is a shocking reality; therefore, users need to be forced, via password management software, to use strong passwords. In addition, passwords should never be stored in plaintext. They should be salted (add a random string of characters prior to hashing), hashed, encrypted, and protected in a password database (Brading 2012). If password management software is not available to force users to create strong passwords, there are some simple password creation tips that may be effective to share with the users, such as using eight characters or more, not using personal information, using a mix of numbers and upper- and lowercase letters, and to avoid dictionary words or those words spelled backward. In addition, encourage users to vary passwords across accounts.

*Another host protection activity is simply updating application software with any and all updates.*

Many cyber-attacks are directed at heavily used applications such as word processors or reader applications. The patch alerts targeted toward users are often ignored. Simply installing security patches can avoid some of the malicious code that, when installed on a host machine, can spread spam, steal data, or take control of the host. NIST (Grance et al., 2004) suggests that organizations should implement a patch management program to assist system administrators in identifying, acquiring, testing, and deploying patches. In addition, a good practice is archiving known vulnerabilities, patches, or resolutions of past problems (Brownlee and Guttman 1998).

*What data are critical and thus should be backed up and secured? Which files are critical and therefore need a cryptographic check sum recorded? When should backups and check sum updates be scheduled?*

A cryptographic checksum is a hash value produced by running an algorithm on a particular file. Essentially, all of the bits of data in a particular document or file are added up and a number or "hash value" is created. This hash value can then be compared to the hash value generated from the same file on another person's computer or at a previous time on the same computer. Preparation should include determining which files are critical and thus need some form of authentication signature, such as a checksum. These values need to be backed up and secured along with the files. If updates to these critical files occur, a new checksum needs to be produced.

*Who will lead the host-based security effort? Where are the host computers located? Who needs to be educated about host-based security? Who has access to the hosts?*

Whoever is leading the host-based security effort also needs to determine and document where the host computers are located (discussed in risk section) as well as who has access to the host computers. The host-based security lead should also facilitate an education program for host-based security.

*How can data be protected while using cloud services?*

Cloud services can be a risk. Sophos (2013) suggests the following steps to protect data on the cloud:

1. Apply URL filtering as part of the web-based policy. This will prevent use of public cloud storage websites as well as other sites deemed prohibited.
2. Control access to particular applications.
3. Files should be automatically encrypted prior to uploading to any cloud storage service. This encryption will be instrumental to the safety of your files if the cloud provider has a security breach.

Not all threats come from the outside and certainly not all of the system vulnerabilities are from your network. If a company has a trade secret such as a process, formula, or a software application that needs to be protected, that company needs to consider how to protect that asset from employee theft. A 2012 survey from Cyber-Ark Software interviewed 820 IT professionals in an attempt to identify security trends. The survey revealed that 55% of the respondents feel their intellectual property has been obtained by their competitors. In addition, nearly half of the respondents indicated that they would take something with them if they knew they were going to be fired tomorrow, such as an e-mail server admin account, financial reports, customer database, and privileged password list. What if the person knew he or she was quitting? Should the person stop attending strategic meetings? Not open sensitive files? YES—especially if he or she is one of seven people in the world who know a trade secret such as a recipe to an extremely popular baked product. This happened to Bimbo Bakeries, the maker of Thomas's English muffins. A top executive took a job offer from a competitor; however, prior to his exit, he accessed sensitive files such as cost-cutting strategies and labor contracts—information that could be damaging to Bimbo in the hands of a competitor (Dale 2010). Thus, this situation ended up in a trade secret fight. The executive was ordered not to work for the competitor, prohibited from divulging Bimbo's trade secrets, ordered to return confidential information, and ordered to notify Bimbo if he accepted employment in the baking industry (Lewis and Roca, LLP 2011).

## 4.4  NETWORK PREPARATION

*Why does the network need continuous monitoring? **Who** will manage this effort?*

It is obvious why we need to monitor our networks continuously, but it is absolutely worth repeating. Refer back to Figure 1.1, which shows the increase of attack sophistication and the decrease of attacker knowledge needed to be destructive. The FBI reports that early in 2012 hackers embedded malicious software in *two* million computers. Accordingly, our networks need to be prepared to protect our data privacy, integrity, and availability. This section will cover some of the tasks a security professional can perform in order to improve the security profile of a network.

*What activities can protect the network?*

*Encrypt network traffic.* All data traveling across a network should be encrypted. For example, e-mail should have the content encrypted as well as all attachments to ensure their integrity. These e-mails are stored on multiple servers, included on backups, and inspected by firewalls. Traffic sniffers should also monitor where these e-mails might be stored or travel.

*Vulnerability management.* There are many aspects to analyzing the vulnerabilities of the network. Nearly all incidents involving vulnerability exploits could have been avoided (NTT 2009). Generally, vulnerability scans are performed on the system to determine where it is weak. Also, companies need to be aware that any open ports on the firewall are exposing the system to the outside world. Mobile devices such as USB drives and phones are also a risk. Applications need to be tested since hackers are now exploiting those as well as the operating system in which they are running.

*Internal risks.* In addition to external risks outside the firewall such as malware and hackers, companies also need to be concerned with internal breaches to their network. There are applications to monitor sensitive data as well as a specific user status. It is also important to identify which groups of users on the network have access to which types of information (National Security Agency 2009).

*Network synchronization—an accurate determination of **when** events occur.* With respect to forensic investigations, timing is very important. In a digital investigation, having the networks in sync will ease the investigators in determining when everything happened. Many organizations use a publically available time server or install a time server behind the firewall. A downside of a public time server is that it leaves a hole in the firewall. A private time server is expensive. When using a cloud configuration, synchronized time becomes even more imperative. The forensics performed on a cloud configuration will be easier to defend if the time stamps from the client-side log files match the time stamps on the provider-side log files (Zimmerman and Glavach 2011).

*Install intrusion detection and prevention systems (IDS/IPS), firewalls, and require authentication.* The network perimeter needs to be configured so that it denies activities not expressly permitted, such as any violation of the company's security policies. To secure the network perimeter, firewalls, IDS, and/or IPSs are typically employed. Depending on the settings, a firewall can block or allow traffic coming from an Internet or network. The drawbacks of a firewall are that the firewall cannot prevent attacks from the intranet, the firewall policy is not dynamic in that it cannot change depending on the attack, and the firewall rules are too simplistic to prevent a virus (Zhang, Li, and Zheng 2004). Thus, in addition to a firewall, an effective network security plan may use an IDS/IPS solution. An IDS monitors the network by looking for any signs of a possible incident. However, the alert will sound after the problem has passed through. An IPS can do the work of an IDS but it can also attempt to stop an incident by preventing access to the network. The NIST "Guide to Intrusion Detection and Prevention Systems" (SP800–94) contains extensive information on IDS/IPS technologies as well as implementation recommendations. Essentially, requiring authentication and installing firewalls and intrusion protection systems should secure network connection points to the organization. The goal is to prevent intrusions, viruses, and zero-day attacks (brand new vulnerabilities that no one knew about) based on rules and packet inspection (Scarfone & Mell, 2007).

*Where did the incident occur?*

As mentioned in the past two sections, knowledge of the network topology will assist in determining the location of the affected systems and servers when an incident occurs. An effective way to document the network topology is by using a blog or bulletin board to notify administrators of changes as well as a wiki to document the network topology (National Security Agency 2009). In this case, however, it is important that the actual network map is in a secure location. Some other information resources that should be included are commonly used ports, operating system documentation, baselines of network and application activity, and hashes of critical files (Grance et al. 2004).

*How to prepare for an advanced persistent threat (APT):*

The way to be prepared for an APT is first to know how to diagnose an APT. An APT is an attack where hackers infiltrate the corporate network and steal sensitive data over a long period of time. These types of hackers are in it for more than the quick buck they could make through identity theft attacks. They are there to steal big secrets, which may take a considerable amount of time. Roger Grimes (2012), a security advisor at InfoWorld, describes an APT as attackers who "exploit dozens to hundreds of computers, logon accounts, and e-mail users, searching for new data and ideas over an extended period of months and years." Traditional methods of attack are

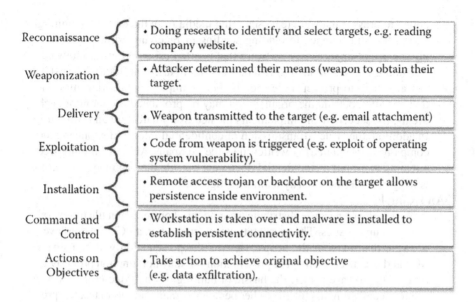

| | |
|---|---|
| Reconnaissance | • Doing research to identify and select targets, e.g. reading company website. |
| Weaponization | • Attacker determined their means (weapon to obtain their target. |
| Delivery | • Weapon transmitted to the target (e.g. email attachment) |
| Exploitation | • Code from weapon is triggered (e.g. exploit of operating system vulnerability). |
| Installation | • Remote access trojan or backdoor on the target allows persistence inside environment. |
| Command and Control | • Workstation is taken over and malware is installed to establish persistent connectivity. |
| Actions on Objectives | • Take action to achieve original objective (e.g. data exfiltration). |

**FIGURE 4.4** Kill chain stages.

often mistaken for APTs. The five signs that you've been hit with an APT are (1) an increase in elevated logons at night (hackers may live in another time zone), (2) widespread backdoor Trojans (a malicious application), (3) unexpected information flows (which can be from one inside computer to another), (4) unexpected data bundles (all stolen data are bundled prior to moving them to the hacker's desired location), and (5) hacking tools that were left behind (such as tools to steal hash numbers) (Grimes 2010).

The next step is to use a defense model that takes into consideration the process the APT attackers use. A team at Lockheed Martin created a model to deal with APT threats. They describe the stages (Figure 4.4) of the attack as a "kill chain." The idea is to use an approach to "stalk the kill chain" by watching the network and being able to identify the kill chain events (Cox 2012). In other words, watch the network as a whole instead of looking at events in isolation.

## 4.5 ESTABLISHING APPROPRIATE POLICIES AND PROCEDURES

*Develop and update an acceptable use policy (AUP).*

*__What__ is an AUP? __Why__ do we need it? __Who__ is affected by the AUP? __How__ do we determine what the __AUP__ should address? __When__ and how often should it be updated? __Where__ should the AUP document be kept?*

An AUP is a document containing an extensive set of rules that restrict how the network and resources may be used. A company can reduce the risk of litigation by publishing and maintaining corporate policies that outline the acceptable use of company resources. An AUP should also outline the policy to prevent malware (e.g., scanning media from the outside, scanning e-mail file attachments, forbidding sending or receiving.exe files, restricting the use of unnecessary software that is often used to transfer malware, restricting removable media) (Mell, Kent, and Nusbaum 2005). The AUP can also help to protect trade secrets. For example, confidential information e-mailed outside the organization may be prohibited. To monitor such events, the security team implements the necessary hardware and software. In addition, the "rule" (transmission of confidential information outside the company is prohibited) is written into the company AUP. If an employee attempts to e-mail a confidential document outside the company, an alarm is triggered and the security team investigates.

**An Example**

An employee needs to distribute a confidential presentation file to several of the company sites. The presentation was created on a PC. The employee is aware that a few of the people viewing the presentation will be using a Mac and is concerned that the presentation file may not function properly in a Mac OS. To save herself the hassle of hunting down a Mac user at work, she innocently e-mails the file to her personal e-mail address to test the presentation on her personal Mac computer. The e-mail was sent, the firewall was tripped, and within minutes she gets the following e-mail:

**From: Mail Delivery Subsystem <MAILER-DAEMON@<companyname>.com**
**Date: Friday, March 1, 2013 5:06 PM**
**Subject: Returned mail: see transcript for details**
Dear Sender:

We have blocked your request to send this e-mail outside the company's firewall.

Please be aware that transmitting this type of content outside the company without prior express authorization may be a violation of company policy.

Do not attempt to retransmit this information in any other form. Sending <company name> confidential information outside the company may violate our policy and may subject you to discipline, up to and including termination of employment.

If you think you have received this message in error, please forward the content of your original message, including who you are sending it to, along with a brief explanation of why your message needs to be sent outside the company to emailprt@<companyname.com>. Your request will be reviewed in a timely manner.

Please contact us with any questions

Regards, <Company Name> Information Protection Services

Figure 4.5 shows an excerpt of an example AUP provided by Sophos, an IT security and data protection provider.

Once the AUP is created and available (usually on the company intranet), it needs to be updated and enforced. The company should require that the AUP be read by all new hires and require a signature prior to using any computer resources at the company. It may also be effective to have the AUP reread annually, as it should be updated and reflect the current trends in technology.

For example, many companies needed to address the bring-your-own-device (BYOD) trend in their AUPs. Just as in anything, there are pros and cons with BYOD.

> ### *Example Policy*
>
> #### 1. Introduction
>
> This acceptable use policy (AUP) for IT Systems is designed to protect <Company X>, our employees, customers, and other partners from harm caused by the misuse of our IT systems and our data. Misuse includes both deliberate and inadvertent actions.
>
> The repercussions of misuse of our systems can be severe. Potential damage includes, but is not limited to, malware infection (e.g., computer viruses), legal and financial penalties for data leakage, and lost productivity resulting from network downtime.
>
> Everyone who works at <Company X> is responsible for the security of our IT systems and the data on them. As such, all employees must ensure they adhere to the guidelines in this policy at all times. Should any employee be unclear on the policy or how it impacts his or her role, he or she should speak to his or her manager or IT security officer.
>
> #### 2. Definitions
>
> "Users" are everyone who has access to any of <Company X>'s IT systems. This includes permanent employees and also temporary employees, contractors, agencies, consultants, suppliers, customers, and business partners.
>
> "Systems" means all IT equipment that connects to the corporate network or accesses corporate applications. This includes, but is not limited to, desktop computers, laptops, smartphones, tablets, printers, data and voice networks, networked devices, software, electronically stored data, portable data storage devices, third-party networking services, telephone handsets, video conferencing systems, and all other similar items commonly understood to be covered by this term.
>
> #### 3. Scope
>
> This is a universal policy that applies to all users and all systems. For some users and/or some systems, a more specific policy exists. In such cases, the more specific policy has precedence in areas where they conflict, but otherwise both policies apply on all other points.
>
> This policy covers only internal use of <Company X>'s systems and does not cover use of our products or services by customers or other third parties.
>
> Some aspects of this policy affect areas governed by local legislation in certain countries (e.g., employee privacy laws). In such cases, the need for local legal compliance has clear precedence over this policy within the bounds of that jurisdiction. In such cases, local teams should develop and issue users with a clarification of how the policy applies locally.
>
> Staff members at <Company X> who monitor and enforce compliance with this policy are responsible for ensuring that they remain compliant with relevant local legislation at all times.
>
> #### 4. Use of IT Systems
>
> All data stored on <Company X>'s systems is the property of <Company X>. Users should be aware that the company cannot guarantee the confidentiality of information stored on any <Company X> system except where required to do so by local laws.
>
> <Company X>'s systems exist to support and enable the business. A small amount of personal use is, in most cases, allowed. However, it must not be in any way detrimental to users' own or their colleagues' productivity, nor should it result in any direct costs being borne by <Company X> other than for trivial amounts (e.g., an occasional short telephone call).

**FIGURE 4.5** Sophos example IT AUP (http://www.sophos.com).

*(Continued)*

<Company X> trusts employees to be fair and sensible when judging what constitutes an acceptable level of personal use of the company's IT systems. If employees are uncertain, they should consult their manager.

Any information that is particularly sensitive or vulnerable must be encrypted and/or securely stored so that unauthorized access is prevented (or at least made extremely difficult). However, this must be done in a way that does not prevent—or risk preventing—legitimate access by all properly authorized parties.

<Company X> can monitor the use of its IT systems and the data on it at any time. This may include (except where precluded by local privacy laws) examination of the content stored within the e-mail and data files of any user, and examination of the access history of any users.

<Company X> reserves the right to audit networks and systems regularly to ensure compliance with this policy.

### 5. Data Security

If data on <Company X>'s systems is classified as confidential, this should be clearly indicated within the data and/or the user interface of the system used to access it. Users must take all necessary steps to prevent unauthorized access to confidential information.

Users are expected to exercise reasonable personal judgment when deciding which information is confidential.

Users must not send, upload, remove on portable media, or otherwise transfer to a non-<Company X> system any information that is designated as confidential, or that they should reasonably regard as being confidential to <Company X>, except where explicitly authorized to do so in the performance of their regular duties.

Users must keep passwords secure and not allow others to access their accounts. Users must ensure all passwords comply with <Company X>'s safe password policy.

Users who are supplied with computer equipment by <Company X> are responsible for the safety and care of that equipment and the security of software and data stored on it and on other <Company X> systems that they can access remotely using it.

Because information on portable devices, such as laptops, tablets, and smartphones, is especially vulnerable, special care should be exercised with these devices: Sensitive information should be stored in encrypted folders only. Users will be held responsible for the consequences of theft of or disclosure of information on portable systems entrusted to their care if they have not taken reasonable precautions to secure it.

All workstations (desktops and laptops) should be secured with a lock-on-idle policy active after, at most, 10 minutes of inactivity. In addition, the screen and keyboard should be manually locked by the responsible user whenever leaving the machine unattended.

Users who have been charged with the management of those systems are responsible for ensuring that they are at all times properly protected against known threats and vulnerabilities as far as is reasonably practicable and compatible with the designated purpose of those systems.

Users must, at all times, guard against the risk of malware (e.g., viruses, spyware, Trojan horses, rootkits, worms, backdoors) being imported into <Company X>'s systems by whatever means and must report any actual or suspected malware infection immediately.

**FIGURE 4.5 (*CONTINUED*)**   Sophos example IT AUP (http://www.sophos.com).

(*Continued*)

**6. Unacceptable Use**

All employees should use their own judgment regarding what is unacceptable use of <Company X>'s systems. The activities below are provided as examples of unacceptable use; however, the list is not exhaustive. Should an employee need to contravene these guidelines in order to perform his or her role, the employee should consult with and obtain approval from his or her manager before proceeding:

- All illegal activities. These include theft, computer hacking, malware distribution, contravening copyrights and patents, and using illegal or unlicensed software or services. These also include activities that contravene data protection regulations.
- All activities detrimental to the success of <Company X>. These include sharing sensitive information outside the company, such as research and development information and customer lists, as well as defamation of the company.
- All activities only for personal benefit that have a negative impact on the day-to-day functioning of the business. These include activities that slow down the computer network (e.g., streaming video, playing networked video games).
- All activities that are inappropriate for <Company X> to be associated with and/or are detrimental to the company's reputation. These include pornography, gambling, inciting hate, bullying, and harassment.
- Circumventing the IT security systems and protocols that <Company X> has put in place.

**7. Enforcement**

<Company X> will not tolerate any misuse of its systems and will discipline anyone found to have contravened the policy, including not exercising reasonable judgment regarding acceptable use. While each situation will be judged on a case-by-case basis, employees should be aware that consequences may include the termination of their employment.

Use of any <Company X>'s resources for any illegal activity will usually be ground for summary dismissal, and <Company X> will not hesitate to cooperate with any criminal investigation and prosecution that may result from such activity.

**FIGURE 4.5 (CONTINUED)** Sophos example IT AUP (http://www.sophos.com).

If not addressed properly, it can be an issue for both the employee and the employer. Here are the pros and cons of BYOD (Bradley 2011):

*Pros:* The users are incurring the cost of the device and the associated expenses (e.g., voice and data). The satisfaction of the users will also increase because they are choosing the brand and model of the device they are bringing to work. They are also updating their devices to the latest and greatest technology much sooner than the organization would.

*Cons:* The company is losing control over the hardware and how the devices are utilized. The AUP and compliance mandates are more difficult to enforce since the device is not company owned. There is also the issue of when the employee separates from the company: The company wants its data back! This will get very tricky if there is a lawsuit involved; the employee may think that his or her phone does not need to be examined, but that is probably not the case. This risk needs to be explained to employees participating in a BYOD program.

Mobile phones are the most popular device to bring. Most mobile devices, especially user-owned mobile devices, can be untrustworthy and do not have many of the trust features that are built into hosts and laptops (Souppaya and Scarfone 2012). NIST (Souppaya & Scarfone, 2012) suggests running the organization's software in a secure state (in isolation from the rest of the device's applications and data) on the mobile phone or utilizing integrity scanning applications for that specific device. Important factors when developing a mobile device security policy include: sensitivity of the work, level of confidence in security policy compliance, cost, work location, technical limitations (if the device needs to be able to run a particular application), and compliance with mandates and other policies (Souppaya and Scarfone 2012).

Other things to consider when writing a BYOD policy (Hassell 2012) include:

1. **Specify what devices are permitted** and supported by the organization.
2. **Establish a stringent security policy for all devices,** such as a complex password attached to the device.
3. **Define a clear service policy for devices under BYOD criteria,** such as support for applications, broken devices, and initial connections.
4. **Make it clear who owns which applications and data** because if the device becomes lost or stolen, the company may wipe the device for security reasons—thus, personal data will be gone. Possibly provide information to the employee regarding backing up his or her own content.
5. **Decide which applications will be allowed or banned.** More than a few mobile device applications have vulnerabilities.
6. **Integrate your BYOD plan with your AUP.** Address social networking, viewing objectionable websites, and how its use will be monitored while a device is connected to the corporate network.
7. **Set up an employee exit strategy** to address how access tokens, e-mail access, data, and proprietary applications and information will be removed.

## 4.6 ESTABLISHING AN INCIDENT RESPONSE TEAM

*Why do we need to establish a response team?*

There is no question that computer incidents are on the rise and that organizations need all the help they can get to be prepared. The computer incident response team (CIRT) is the much needed point of contact when an incident occurs. The CIRT can help determine the incident's impact on the organization as well as perform tasks that can mitigate the damage and get the organization up and running again. They can also help with risk assessment, offer lessons learned, and help with training.

*Who should be on the CIRT?*

Depending on the needs of the organization, the team can consist of employees, be fully outsourced, or be a combination of the two. This choice will depend on the organization's resources and needs.

NIST has written a "Computer Security Incident Handling Guide" (Cichonski et al. 2012) that addresses the CIRT. It contains CIRT recommendations that address *what* factors need to be considered, *how* to establish the team, *when* to contact stakeholders, etc.:

- Establish a formal IRincident response capability to be prepared when there is a breach.
- **Create an IRincident response policy** to define "incidents," roles, and responsibilities, etc.
- **Develop an IRincident response plan based on the IRincident response policy.** Also, establish metrics to assess the program and how training should occur.
- **Develop IRincident response procedures** that detail the step-by-step process of responding to an incident.
- **Establish policies and procedures regarding incident-related information sharing** such as media, and law enforcement, etc. The team needs to consult with the organization's legal department, public affairs, and management to determine the policy and procedure.
- **Provide pertinent information on incidents to the appropriate organization,** such as the US-CERT for federal civilian agencies and/or ISAC organizations who use data to report threats and incidents.
- **Consider the relevant factors when selecting an IRincident response team model** to construct the most effective team structure (e.g., to outsource or not) for the organization.
- **Select people with appropriate skills for the IRincident response team.** In addition to technical skills, being able to effectively communicate should be a requirement.
- Identify other groups within the organization that may need to participate in incident handling, such as management, legal, and facilities, etc.
- **Determine which services the team should offer** in addition to IRincident response, such as monitoring intrusion detection sensors, disseminating information regarding security threats, and educating users on the security policies.

In addition to the preceding, are a few other elements to consider when building an effective CIRT team:

*Train the response team (what).*
 Given the technology growth rate, the CIRT team will probably need additional training. A properly trained team is as important as having a secure network, as it will increase the chances of data validity upon collection. In general, the response team is responsible for facilitating technical assistance (analyzing compromised systems), eradication (elimination of the cause and effect of incidents), and recovery (restoring systems and services) (Brownlee and Guttman 1998). Outside training involving certifications (as discussed in Chapter 2) is an option to educate the team effectively on procedure, process, and documentation. A major part of this training should be learning how to ensure the integrity of the evidence. This will be discussed in the next chapter.

*Maintain the chain of custody (where).*

Another task for the CIRT is to maintain the chain of custody of the incident evidence collected. The location of the evidence from the moment it was collected to the moment it is presented in court needs to be traceable (Mandia et al. 2003). Chain-of-custody validation is one way in which a court can verify the authenticity of electronic evidence. Items to be documented include (Brezinski and Killalea 2002):

- Where, when, and by whom was the evidence collected?
- Where, when, and by whom was the evidence handled or analyzed?
- Who had custody during what period of time? How was evidence stored?
- If evidence changed custody, document when and how the transfer of custody occurred. Include all shipping information.

## 4.7 PREPARING A RESPONSE TOOLKIT

*When and why is a response toolkit necessary? Where is it used?*

A response toolkit is no different from the bag of tools an emergency medical responder carries or the bag a parent carries on an airplane when traveling with a young child. The toolkit prepares one for when an emergency occurs. Specifically, the IR toolkit will help acquire evidence and protect it from contamination. It is a mobile toolkit to be used wherever the problem occurs, and it is contained on mobile media such as disks or USB drives. Being prepared—that is, having all of the tools needed to deal with a breach—will reduce the damage and downtime in the organization as well as the efficiency and effectiveness of the investigation.

*How is the toolkit assembled?*

The toolkit is assembled with tools that are needed in an incident including both hardware and software. Some of the software tools are system utilities (netstat, mkdir, find, etc.), and some tools are needed to analyze data such as EnCase or FTK. The toolkit needs to have the system utilities that are already on the machine because some hackers will install their own versions of system utilities (e.g., installing a new mkdir that contains malware) on your system. This fact stresses the importance of having forensically sound (in good condition) tools available in an emergency.

Acquire necessary software to respond to incidents. Acquire necessary hardware to respond to incidents (**what**).

Carlton and Worthley have demonstrated the importance of preparing a response toolkit (2009). They collected data from both computer forensic examiners and attorneys with computer forensic experience and determined that one of the top data acquisition tasks agreed upon by both was "wiping target disk drives and verifying target disk drives are wiped." Thus, some of the main software tools besides system utilities needed in the toolkit are tools that will image the drive (e.g., dd, EnCase) and will be able to analyze/search the data (e.g., EnCase, FTK, Paraben).

Carlton and Worthley's results also showed agreement on the following tasks:

| Computer Security Incident Report |
|---|

**Date and time:**
Location:
Status:

| **Response lead:** | **Reported by:** |
|---|---|
| Name: | Name: |
| Job title: | Job title: |
| Phone : | Phone : |
| Mobile: | Mobile: |
| Email: | Email: |

**Type of incident**

☐ Exposing confidential/classified/unclassified data    ☐ Fraud
☐ Denial of service                                             ☐ Destroying data
☐ Malware                                                   ☐ Unauthorized use
☐ Unauthorized access                              ☐ Other
Comments:

**Description of incident**

Date/time:                         Date: _____           Time: _____
Has the attack ended?         ☐ Yes        ☐ No
Duration of attack (in hours):    ___
Severity of attack:              ☐ Low       ☐ Med       ☐ High

**Actions taken**

Identification measures:
Containment measures:
Evidence collected:
Eradication measures:
Summary of incident:
Cause of incident:
Damage of incident:

**Postmortem**

What worked well:

Lessons learned:

Procedure corrections that would have improved recovery efforts:

**FIGURE 4.6**  Incident response template.

- Prepare and verify toolkit to ensure that equipment is fully functional.
- Test forensic software tools.
- Prepare and verify toolkit to ensure that all necessary hardware connectors and adapters are fully stocked.

*Acquire necessary documentation to respond to incidents (what).*

Documentation needs to be standardized to ensure that all necessary items are recorded. Of the top 26 data acquisition tasks that resulted from Carlton and Worthley's work, ten involved documenting specific portions of the investigation. NIST (Cichonski et. al., 2012) suggests utilizing an issue tracking system to document the following:

- Status of the incident
- Incident summary
- Any indicators related to the incident
- All actions taken by any incident handlers
- Chain of custody
- Impact on the organization
- Contact information for all stakeholders
- A list of all evidence acquired during the investigation
- Comments
- Next steps

There are many variations of incident forms available on the web. Figure 4.6 shows an example IR form to be filled out when reporting an incident.

## 4.8   TRAINING

*Why is it important to educate the users? Who will do the training? How often and where should the training occur (when) (where)?*

W. Edwards Deming, an electrical engineer well known for his management teachings and philosophy and most famous for his creation of 14 points to facilitate the improvement of processes, systems, products, and services, encouraged "training on the job." This was point number 6 of the famous 14. His point was that training is required to be able to know and understand the new skills that are required to do a job. Thus, the security strategy of all companies needs to include a *user training* layer in addition to firewalls and intrusion detection, etc. because no single person or team is responsible for the security of an organization. Everyone needs to play a role in protecting the critical assets and, furthermore, everyone is a weakness. As a matter of fact, a major security issue an enterprise faces is precarious user behavior. For example, ask a group of people what they would do if they found a USB drive in the parking lot. Most likely, at least one of them will tell you that he or she would insert the drive into a computer to determine the owner, consequently failing the penetration test. A penetration test is a method to evaluate the security of a system or network by exploiting a vulnerability. The found USB drive should be treated like a used tissue on the ground because both probably carry a virus. In another penetration test example, Symantec placed 50 smartphones around five major cities and discovered that when people found the mobile phones they accessed the users' sensitive corporate data, contact information, cloud-based documents, social networking, passwords, salary information, and online banking (Vance 2012). It is really a wake-up call for employers to train their employees to use strong passwords for certain applications and a PIN to unlock the phone. In addition, the stolen/found mobile phones are also being sold on black markets with each credential recovered off the phone an added bonus in the amount the phone is worth. In these examples, the weakness is the user, indicating that security training needs to be renewed. Penetration testing reminds me of the Russian proverb that Ronald Reagan made famous: **"Trust but verify."** You may trust your employees but the training and penetration testing will help to verify they know the right thing to do if confronted by a situation that could put the security of your business at risk.

Security training can be accomplished via off-site training, in-house classroom training, security-awareness websites, pushing out tips at start-up, periodically e-mailing security tips, or decorating the office with various safety reminder posters (SANS Institute 2009). There is obviously no guarantee that training will remove the problem of users clicking on an infected attachment or giving away their passwords, but training needs to be part of a comprehensive security profile to reduce the risk.

*Educate users about proper use and malware. Use lessons learned, penetration testing, and live testing (how) (what).*

Users should be informed about the appropriate use of networks, hosts, and the applications they use. Training should also include guidance about malware incident prevention, which can mitigate malware incidents (Brownlee and Guttman 1998). This goal can be accomplished by sharing "lessons learned" from previous incidents so that they can see how their actions could affect the organization (Stoneburner et al. 2002). As described earlier, training can also be a result of penetration testing. Another example of a penetration test is to send inappropriate "test" e-mails to employees with the goal of education. For example, one could simulate a phishing/social engineering attack by sending out an e-mail asking users for their username and password to see how many would actually send that information back.

In addition to understanding appropriate use, users should know how to contact the response team as well as understand the services they provide. The Network Working Group suggests publishing a clear statement of the policies and procedures of the response team in order for the users to understand how to report incidents and what to expect after an incident is reported (Zimmerman and Glavach 2011).

The IT staff also needs to be trained to be able to maintain the hosts, networks, and applications in accordance with the security standards of the organization (Mandia et al. 2003). One training option is *live testing.* An example of live testing could entail simulating a cyber security incident and evaluating the reaction and processes of the IR team. This technique is often used in educational settings.

The following is a simplified view of the *priority 1* recommendations by NIST for the **awareness and training baseline:**

|  | Impact Level | | |
| --- | --- | --- | --- |
| Control Name | Low | Moderate | High |
| Formal documentation to facilitate the implementation of the security awareness and training policy and procedure | ✓ | ✓ | ✓ |
| Provide security-awareness training for new employees and when changes occur in the system | ✓ | ✓ | ✓ |
| Provide security training covering technical, physical, and personal safeguards and countermeasures required prior to access | ✓ | ✓ | ✓ |

## THE TOP TEN WAYS TO SHUT DOWN THE NO-TECH HACKER

1. *Go undercover:* Be a little paranoid. Some hackers are looking for the company logo on your PC while you are working at the coffee shop, or waiting for you to discuss company secrets at the lunch hangout near your office. So cover up the company logos and keep the conversation light.
2. *Shred everything:* Some laws require the proper disposal of private information (HIPAA). There are people that may look through your trash hunting for personal information as well. If you do not have a shredder, you can use scissors, burn the documents outside in an open area, or submerge papers in water overnight.
3. *Get decent locks:* Install or use the locks that the professionals recommend—the locks that cannot be tampered with easily. It is also recommended that the keys be hidden.
4. *Put that badge away:* If a hacker gets one look at your badge, he or she will probably have no problem duplicating it.
5. *Check your surveillance gear:* Install quality cameras to minimize tampering, use multiple cameras for the same view, protect the camera from physical attack with housing, and consider hidden cameras.
6. *Shut down shoulder surfers:* No-tech hackers also like to watch what you are working on from afar (or over your shoulder). If you

are working on something sensitive, be cognizant of your angle (e.g., sit with your back against the wall). When punching in pass codes, shield with your hand. If you suspect that someone is watching, stop what you are doing, close your screen, and determine if anything sensitive was compromised.

7. *Block tailgaters:* This is referring to people that walk in behind you after you have been cleared for entrance. Do not let them in! Challenge people you cannot identify and/or notify security.

8. *Clean your car:* Stickers on your car (e.g., parking permits) and personal papers in your car give away a lot of information.

9. *Watch your back online:* Never enter your personal information in an instant messenger or web browser.

10. *Beware of social engineers:* They are eliciting sensitive information from you. See Chapter 1 for more on social engineering.

*Taken from Long (2008).*

## REFERENCES

Brading, A. 2012. Yahoo Voices hacked, nearly half a million emails and passwords stolen. naked security.sophos.com (retrieved July 12, 2012).

Bradley, T. 2011. Pros and cons of bringing your own device to work. *PCWorld* (retrieved December 20, 2011).

Brezinski, D. and Killalea, T. February 2002. Guidelines for evidence collection and archiving. Network Working Group RFC 3227.

Brownlee, N. and Guttman, E. June 1998. Expectations for computer security incident response. Network Working Group RFC 2350.

Carlton, G. and Worthley, R. 2009. An evaluation of agreement and conflict among computer forensics experts. *42nd Hawaii International Conference on System Sciences*, pp. 1–10.

Cichonski, P., Millar, T., Grance, T. and Scarfone, K. August 2012. Computer security incident handling guide. NIST special publication 800–61, revision 2.

Cluley, G. 2012. The worst passwords you could ever choose exposed by Yahoo Voices hack. Sophos nakedsecurity.sophos.com (retrieved July 13, 2012).

Cox, A. August 16, 2012. Stalking the kill chain: The attacker's chain. *RSA FirstWatch.*

Cyber-Ark. 2012. 2012 Trust, security & passwords survey. http://www.websecure.com.au/blog/2012/06/cyber-ark-2012-trust-security-and-passwords-survey.

Dale, M. July 29, 2010. Secret of English muffin "nooks & crannies" is safe for now. *USA Today.*

DeFranco, J. and Laplante, P. 2011. Preparing for incident response using the Zachman framework. *IA Newsletter* 14(3).

Grance, T., Kent, K. and Kim, B. 2004. Computer security incident handling guide. National Institute of Standards and Technology, special publication 800–61. http://www.csrc.nist.gov/publications/nistpubs/800-61/sp800-61.pdf (retrieved January 19, 2010).

Grimes, R. October 19, 2010. How advanced persistent threats bypass your network security. *InfoWorld.*

Grimes, R. October 16, 2012. 5 signs you've been hit with an advanced persistent threat. *InfoWorld.*

Hassell, J. May 17, 2012. 7 Tips for establishing a successful BYOD policy. *CIO Magazine.*

Leong, R. 2006. FORZA—Digital forensics investigation framework that incorporates legal issues. *Digital Investigation* 3S: 29–36.

Lewis and Roca, LLP. 2011. Protecting "nooks and crannies" *Bimbo Bakeries USA, INC. V. Chris Botticella.* http://www.lrlaw.com/ipblog/blog.aspx?entry=260 (retrieved January 14, 2013).

Long, J. 2008. *No tech hacking: A guide to social engineering, dumpster diving, and shoulder surfing.* Burlington, MA: Syngress Press.

Lyle, J., Guttman, B., Butler, J., Sauerwein, K., Reed, C., Lloyd, C. May 2022. Digital investigation techniques: A NIST scientific foundation review. https://nvlpubs.nist.gov/nistpubs/ir/2022/NIST.IR.8354-draft.pdf (retrieved May 12, 2022).

Mandia, K., Prosise, C. and Pepe, M. 2003. *Incident response & computer forensics*, 2nd ed. New York: McGraw–Hill.

Mell, P., Kent, K. and Nusbaum, J. November 2005. Guide to malware incident prevention and handling. NIST special publication SP800-83.

National Security Agency. 2009. Manageable network plan.

NTT. September 2009. Communications white paper, 8 elements of complete vulnerability management.

Ponemon Institute. 2012. Cost of cyber crime study: United States. http://www.ponemon.org/local/upload/file/2012_US_Cost_of_Cyber_Crime_Study_FINAL6%20.pdf (retrieved January 10, 2013).

SANS Institute. 2009. The importance of security awareness training. http://www.sans.org/reading_room/whitepapers/awareness/importance-security-awareness-training_33013 (retrieved January 20, 2013).

Scarfone, K. and Mell, P. 2007. Guide to intrusion detection and prevention systems. NIST special publication 800-94.

Sophos. 2013. Security threat report 2013. http://www.sophos.com/en-us/security-news-trends/reports/security-threat-report.aspx (retrieved January 9, 2013).

Souppaya, M. and Scarfone, K. July 2012. Guidelines for managing and securing mobile devices in the enterprise. NIST special publication 800-124.

Stoneburner, G., Goguen, A. and Feringa, A. 2002. Risk management guide for information technology systems. NIST special publication 800-30. http://csrc.nist.gov/publications/nistpubs/800-30/sp800-30.pdf (retrieved on January 23, 2010).

Vance, A. March 8, 2012. Data security: Most finders of lost smartphones are snoops. *Bloomberg Businessweek.*

Zachman, J. 1987. A framework for information systems architecture. *IBM Systems Journal* 26(3): 276–292.

Zhang, X., Li, C. and Zheng, W. 2004. Intrusion prevention system design. *The 4th International Conference on Computer and Information Technology.*

Zimmerman, S. and Glavach, D. 2011. Cyber forensics in the cloud. *IA Newsletter* 14(1).

## NOTES

1 Excerpts in this chapter are from DeFranco and Laplante (2011).
2 National Vulnerability Database: http://web.nvd.nist.gov/view/ncp/repository.

# 5 Incident Response and Digital Forensics

Efficiency is doing things right; effectiveness is doing the right things.

—*Peter F. Drucker*

## 5.1 INTRODUCTION

Incident response (IR) and digital forensics (DF) need both efficiency and effectiveness because if they are not done correctly, your efforts will be futile. In this chapter, the fundamental processes for IR and digital forensic analysis will be discussed. Just today, an *incident* occurred on my laptop, no less. Similarly to every other day, I dock my laptop upon my arrival and start checking my e-mail. Within a few minutes, the IT admin is at my door and announces that we have a problem. He said he received a message from the main IT office—over 300 miles away and monitoring over 20 locations and thousands of computers—that my laptop has been compromised. He was instructed to remove it from the network and begin the analysis process by scanning it for any personal information that may have been accessed by a hacker. This is a great example of an incident; as small as it sounds, it is, in fact, an incident. The official definition of an incident is a situation that has compromised the integrity, confidentiality, or availability of an enterprise network, host, or data. Other incident examples include attempting to gain unauthorized access to a system, a DDOS (distributed denial of service) attack, unauthorized use of a system, and website defacement.

Generally, the IR process is to detect, contain, and eradicate the incident, and the DF process is to collect, analyze, and report the evidence. In other words, once the incident is contained and eradicated, the DF professional begins the evidence collection process. The goal of the analysis is to determine (1) what happened so that reoccurrence of the incident can be avoided, and (2) whether this is a criminal case.

The cases where electronic evidence is critical are not always action packed with computer break-ins, SQL (structured query language) injections, DDOS, malware, phishing attempts, or company web page defacement. Some cases requiring electronic evidence are disloyal employees who are suspected of industrial espionage,[1] breached contracts, an employee dismissal dispute, theft of company documents, inappropriate use of company resources (e.g., possession of pornography), copyright infringement (music illegally traded over the Internet), harassment (e-mail-based stalking), and identity theft.

## 5.2 INCIDENT RESPONSE

Be prepared. We have all heard that before—especially if you were a Boy Scout or if you read the preceding chapter. This is true of life in general as well. For example,

DOI: 10.1201/9781003245223-5

financial experts tell us to prepare for job loss by having 3–6 months of savings available. In a poor economy, we should have more, but the point is that we are taught to prepare for that rainy day. When an incident occurs on your system, it may be more of a hurricane. If you are prepared and monitor anything that could impede success, when something unplanned occurs, you are prepared to deal with the issue in order to get your system back up. It is like being the only house with a generator after the hurricane caused a neighborhood power loss.

The National Institute of Standards and Technology (NIST) has provided a baseline for **IR**. Here is the simplified view of the *priority 1* recommendations. The first control (creating documentation of the IR policy and procedures) and the last control (creating an IR plan) are part of preparing for IR, which were addressed in Chapter 4. The other controls listed will be addressed in this chapter.

IR is a life cycle of stages shown in Figure 5.1. We covered *preparation* in the last chapter (e.g., establishing the computer incident response team [CIRT], training the users, and installing the necessary hardware and software). The next stage, *detection/identification*, is more difficult to address because incidents are not always apparent; hence, constant monitoring (using tools acquired during the prep stage) of the assets is required to detect an incident. If an anomaly is detected, the situation is analyzed to confirm that an incident is occurring. If the incident is confirmed, it needs to be *contained* (so as not to infect other parts of the system) and *eradicated*. And, finally, the *recovery* process will bring the system back to working order.

| Control Name | Impact Level | | |
| --- | --- | --- | --- |
| | Low | Moderate | High |
| Formal documentation of the IR policy and procedures | ✓ | ✓ | ✓ |
| Incident handling capability to include preparation, detection and analysis, containment, eradication, and recovery | ✓ | ✓ | ✓ |
| Implement monitoring and documentation of incidents | ✓ | ✓ | ✓ |
| Require incident reporting within a defined time period | ✓ | ✓ | ✓ |
| IR plan that is a road map for response capability and also describes the structure and organization of the IR capability | ✓ | ✓ | ✓ |

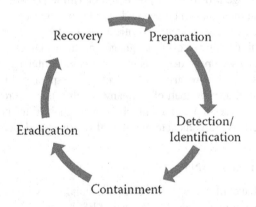

**FIGURE 5.1**    Incident response life cycle.

### 5.2.1 DETECTION/IDENTIFICATION

In this stage, the monitoring has produced an inconsistency or alarm that needs to be investigated to determine if an incident actually occurred. Incidents may also be discovered by a system administrator or even an end user. In any case, the first step is to verify that the "incident" is not actually an error. For example, a user error, a system/software configuration, or a hardware failure could present itself as an incident. Ways to confirm an incident include analyzing the technical details such as reports and logs, interviewing any personnel who may have insight, and reviewing the access control lists of the network topology (Mandia, Prosise, and Pepe 2003).

If it is concluded upon analysis that the incident is not an error, the type of incident needs to be determined. There are two types of incidents: (1) a *precursor* that an incident is imminent, and (2) an *indicator* that the incident is occurring or has occurred (Cichonski et al. 2012). An *incident precursor,* for example, could be server entries of a vulnerability scanner, knowledge of a new mail server exploit, or a directed threat at the organization. Some possible *incident indicators* are, for example, IDPS (intrusion, detection, and prevention systems) or antivirus alerts, a logged configuration change, failed login attempts, a large quantity of bounced e-mails, or unusual network traffic. It is not an easy process to validate an incident since an alert can be a false positive. In order to perform an effective analysis when an incident occurs, NIST (2012) recommends that the following items be in place to determine the scope of the incident (or precursor) more efficiently:

- Have the networks and systems profiled for normal use so file integrity and changes can easily be identified.
- Understand normal behaviors of networks, systems, and applications by reviewing log entries and security alerts so that abnormalities can be easily identified.
- Create a log retention policy to determine how long log data from firewalls, IDPSs, and applications should be stored. Log data are helpful in the analysis of an incident.
- Perform event correlation between all of the available logs (e.g., firewall, IDPS, application) as they all record different aspects of the attack.
- Keep all host clocks synchronized. As was discussed in the last chapter, it is important during an investigation that all of the logs show the same time that an attack occurred.
- Maintain a knowledge base of searchable information related to incidents and the IR process.
- Use a separate work station for web research on unusual activity.
- Run packet sniffers (configured to specified criteria) to collect additional network traffic.
- Have a strategy in place to filter the data on categories of indicators that are of high significance to the organization's situation.
- Have plan B in place. If the incident scope is larger than can be handled by your team, seek assistance from external resources.

With the proliferation of ransomware, malware, and many other attacks that can result in the destruction or alteration of data, it is imperative that businesses create specific IR guidelines—encouraging effective detection.

For more information on detection, NIST (2020) has created a cybersecurity practice guide (SP 1800–11B) with the following goals:

- Restore data to its last known good configuration
- Identify the correct backup version (w/o malicious code with data restoration)
- Identify altered data as well as the date and time of alteration
- Determine the identity/identities of those who alter data
- Identify other events that coincide with data alteration
- Determine any impact of the data alteration

The full publication with extensive guidelines to meet these goals can be found at https://nvlpubs.nist.gov/nistpubs/SpecialPublications/NIST.SP.1800-11.pdf.

Finally, if an incident is a reality and the containment process is started, make sure that any evidence is documented, a chain of custody of any evidence collected is maintained, and the incident is reported to the appropriate officials within a defined time period.

### 5.2.2 CONTAINMENT

The goal of the containment stage is to minimize the scope and damage of the incident. The containment strategy will depend on certain aspects of the incident, such as the damage/theft of resources, the need for evidence preservation, service availability, time and resources available to implement the strategy, and duration of the solution (Cichonski et al. 2012). For example, in a DDOS attack, shown in Figure 5.2, the attacker is attempting to make the resource unavailable to the users by a sending a flood of messages from compromised computers, which the attacker is controlling, to a network. Essentially, it is more traffic than it can handle, which means it will be inaccessible to a legitimate user.

These types of attacks can bring your favorite social networking website to a standstill for hours until the attack on the website stops. One containment strategy for DDOS is filtering the traffic directed at the victim host and then locating the machines doing the attacking. This is obviously more easily said than done because there could be 300–400 unique IP addresses doing the attacking. DDOS attacks are not uncommon. If you think your assets are at risk for a DDOS, then contracting with a DDOS mitigation firm—before the attack occurs, of course—may be a good idea. If you are under attack, there are DDOS mitigation firms that will help, but that is like calling on the heating, ventilation, and air conditioning service when your air conditioner does not work during a heat wave.

Containment strategies for other incidents include maintaining a low profile (so the attacker is not tipped off), avoiding potentially compromised code (recall in the last chapter that some hackers like to install their own versions of system utilities), backing up the system (in the event that evidence is needed), and changing the passwords on any of the compromised systems (FCC 2001).

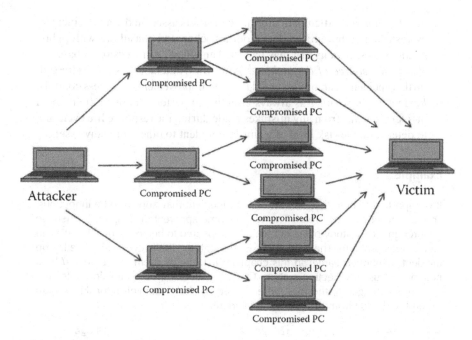

**FIGURE 5.2** DDOS attack.

### 5.2.3 ERADICATION

The goal of the eradication phase is to remove the cause of the incident. In addition to determining the type of attack, the containment phase hopefully provided insight into how the attack was executed—information that may help in determining an effective eradication strategy. For example, eliminating the cause of the incident may involve removing viruses or deactivating accounts that may have been breached as well as securing the vulnerability that facilitated the attack. Therefore, a clean backup to reimage the system will be needed to ensure that malicious content is gone and that any problems cannot be spread. Then, the appropriate protection to secure the system will be implemented.

### 5.2.4 RECOVERY

The goal of the recovery phase is to get the system back up to normal operation. The system should also be validated. Validation means that the system is in fact operating normally after the restoration. The system should also be monitored for reinfection and for anything that was not detected originally. Once the system is back up and running, a follow-up analysis on the incident would be effective. Some aspects of the analysis should be the following (Lynch 2005):

- *Damage assessment:* Determine the systems, networks, and data affected; then identify possible remediation steps.

- *Reverse damage:* Attempt to minimize the costs associated with the incident by restoring compromised data from a backup and consulting with public relations on situations that may have had an effect on any customer base.
- *Nullify the source of the incident:* After the vulnerabilities are addressed, further incidents can be prevented through improvements to access controls.
- *Review the incident:* It is always effective to perform a postmortem on an incident to learn from any mistakes made during the response life cycle and to determine the risk level of a similar incident to other company assets.

### Example

It is important to note that even though an incident may appear to be innocuous, you can never be too careful. This will become apparent in Chapter 7, when we go through a case study of an incident that appeared to be no big deal but was, in fact, a huge problem. That said, let us go through the IR life cycle using the laptop incident mentioned earlier in this chapter. First, the problem was *detected* by a network intrusion detection system. Once the alert from the intrusion detection system was triggered, the logs were viewed by the IT administrator. Here is an excerpt of the log from this particular incident:

```
– – – – – – – -All logs are EST– – – – – – – – - Snort Logs
= = = = = = = = = = = Feb 20 09:28:56 ET TROJAN Backdoor.
Win32.Pushdo.s Checkin
[Classification: A Network Trojan was detected] {TCP}
1XX.1XX.1XX.1XX:58989 -> 213.182.5.:80 Feb 20 09:28:56 ET
TROJAN
Backdoor.Win32.Pushdo.s Chec
kin..........................................................................................................
```

**NOTE:** Backdoor.Win32.Pushdo.s is a Trojan that allows unauthorized access and control of an affected computer.

To *contain* the incident, the laptop was immediately taken off the network (and out of my office). *Eradication* primarily consisted of removing the Trojan; however, the network traffic was also analyzed for any communication of the laptop with any malicious websites, as well as for any network vulnerabilities that were not known prior to the incident. *Recovery* consisted of patching any vulnerabilities that were determined based on log analysis, blocking the IP address of any malicious websites that were communicating with the laptop, and rebuilding the laptop by backing up personal files and reinstalling the OS, user applications, and personal files. Finally, if any personal information was compromised, those individuals would be notified.

## 5.3   INCIDENT RESPONSE FOR CLOUD COMPUTING

If you are thinking about utilizing cloud computing, it is imperative that the IR procedure of the cloud provider be understood before any commitments are made. An organization utilizing cloud computing should consider the following in an IR plan (Jansen and Grance 2011):

## THE INTERNET CRIME COMPLAINT CENTER (IC3)

If you believe you are the victim of an Internet crime, or if you are aware of an attempted Internet crime, you should file a complaint with the IC3, which is an alliance between the National White Collar Crime Center and the Federal Bureau of Investigation (FBI). The IC3's mission is to reduce economic crimes committed over the Internet. Internet crime is defined as:

...any illegal activity involving one or more components of the Internet, such as websites, chat rooms, and/or e-mail. Internet crime involves the use of the Internet to communicate false or fraudulent representations to consumers. These crimes may include, but are not limited to, advance-fee schemes, non-delivery of goods or services, computer hacking, or employment/business opportunity schemes.

The IC3 crime repository not only benefits the consumer, but also helps law enforcement to reduce Internet crime by providing training in being able to identify Internet crime issues. It is also an effective way for different law enforcement and regulatory agencies to share data.

Once a complaint is submitted, the IC3's trained analysts review and research each complaint and then disseminate the information to the appropriate federal, state, local, or international law enforcement agency. To file an Internet crime complaint, visit the IC3 website at http://www.ic3.gov.

- Event data must be available in order to detect an incident. Depending on what type of cloud service is being provided (IaaS [infrastructure as a service], PaaS [platform as a service], SaaS [ software as a service]), event logs may or may not be available to the customer. IaaS customers have the most access to event sources.
- The scope of the incident needs to be determined quickly. It should include a forensic copy of the incident in the event that legal proceedings are necessary.
- Containment will again depend on the cloud service provided. For example, if SaaS is the cloud service, containment may mean taking the software off-line.

Top ten *Threat Actions* (What the Hacker Did to Cause the Breach)

1. Keylogger (spyware that captures data from user activity)
2. Guessing login credentials
3. Stolen login credentials
4. Sending data to external sites
5. Brute force and dictionary attacks (hacking by systematically using an exhaustive word list)

6. Using backdoor malware
7. Hacking the backdoor (gaining unauthorized access to a network)
8. Manipulating the security controls
9. Tampering
10. Exploiting insufficient authentication

(Taken from Verizon's 2012 Data Breach Investigations Report.)

Some of those are easy fixes. Incidentally, Verizon also analyzed dates and locations of incidents and determined that hackers do most of their work Saturday through Monday and that Monday is the most productive. Good to know!

## 5.4   DIGITAL FORENSICS

Crimes occur and the investigations hit a dead end because there appears to be no witness or evidence—that is, until DF comes into play. The FBI solved a case using computer forensics in 2008 where a tip was received regarding two children being sexually abused at a hotel (FBI.gov 2011). Unfortunately, by the time the tip was received, the crime had occurred. It appeared there was no evidence to charge anyone until the computer of the accused was analyzed. The evidence on the computer, a deleted e-mail with directions to the hotel where the abuse occurred, was enough to charge three adults, who are now serving life sentences in prison.

There are also times when you do not know a crime was committed until a forensic analysis is performed. Recall the situation described about the trade secret accessed by an executive after she quit and before she left to work for a competitor. The point is that an incident may appear to be innocuous or impossible to solve until the situation is analyzed. For example, in a situation where the server seemingly went off-line for no reason, after analyzing the log files, you may determine that the cause was malware installed after an intrusion. If a crime has been committed or is even suspected, it is of the utmost importance that the investigator has collected and documented the evidence in a forensically sound manner because the next step would be to hand all of the evidence off to law enforcement.

Another application for DF is evidence gathering for e-discovery—the pretrial phase where electronic evidence is collected. For example, a lawyer may want to prove a spouse's infidelity and may use a forensic analysis of e-mail files to prove the accusation. In evidence gathering, technique and accuracy are critical to ensure the authenticity of the data collected when an incident occurs. The forensic investigator needs always to keep in mind that he or she may be called on to defend the techniques utilized to gather the evidence. In Chapter 6, case law is presented to demonstrate what can happen when the law is not followed while collecting and preserving evidence.

Handling digital evidence is a complex process that should be handled by a professional. If not handled with care, it can be easily destroyed and rendered inadmissible if a court case ensues. There is evidence that can be easily found but other evidence may have been hidden, deleted, or encrypted. Adding to the complexity, if the evidence is not handled properly, it will be thrown out or the

**FIGURE 5.3**   Digital forensics life cycle.

case will be lost. The four main stages of the DF life cycle (Figure 5.3) provide guidance for a forensic expert. The remainder of this section will review the tasks involved in each stage.

### 5.4.1 PREPARATION

Being prepared to perform a forensic investigation involving digital evidence will save a lot of time and effort in the long run. This includes having all of the tools and equipment needed as well as understanding how to use those tools effectively to collect and analyze the evidence. In addition, there should be a plan in place as to where the evidence will be stored securely to prevent contamination or data destruction.

Generally, these are the types of software and hardware needed to perform a forensic investigation: software to duplicate evidence in a forensically sound manner (that does not alter the evidence), software to analyze the drive space that was duplicated (including, at a minimum, features to identify deleted files and filter through keyword searches), and a write blocker to show that the disk copy did not modify the evidence. For example, the write blocker shown in Figure 5.4 is designed to allow forensically sound images on a USB to be extracted without fear that the data on the USB will be modified during the process.

Other items that may be useful during an investigation are notebooks, evidence bags, tape, labels, pens, cameras, antistatic bags (to transport electronic components), Faraday bags (bags to shield a device from signals that may modify evidence), and clean drives (wiped of all other data) to store the duplicated drive image. To wipe a drive, disk sanitation software[2] is used to write zeros over every bit of a drive. This is not an exhaustive list of items, but it gives you an idea of what types of things need to be considered when preparing for a digital forensic investigation.

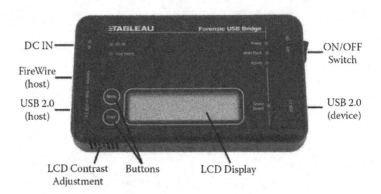

**FIGURE 5.4**   Tableau T-8 Write Blocker from Guidance Software (http://www.tableau.com).

## 5.4.2 COLLECTION

Part of the evidence collection is to document the scene. This includes document-ing things like the model and make of the devices under investigation as well as photographing the surroundings. For example, the investigator may take a photo of the screen to show what was happening when the scene was entered. In addi-tion, the investigator may check out the taskbar and take a photo of the maxi-mized applications running. If the investigator is subpoenaed to come to court, one of the things the lawyer may do is attempt to put some doubt surrounding his or her credibility. The lawyer may ask the color of the door to the office where the computer resided. If the investigator says brown and it was dark blue, that will be strike one.

Once the scene is documented, the electronic data need to be dealt with. In respond-ing to an incident 30 years ago, a forensic investigator would not have thought twice about the "pull the plug" method, which means shutting it down, bringing it back to a lab, and duplicating the hard drive. Due to the increase in complexity in today's computing, the investigator's response is not the same for every incident, so powering it down right away may not be in the best interest of this investigation. (Note: taking it off the network is a good idea to avoid further damage.) The reason that the computer should not be turned off is that the volatile data (e.g., running processes or network connections) are lost. In addition, there is a risk that a rogue application may start a malicious attack when a shutdown is detected. Even the duplication step has changed. Not only have hard drive sizes increased considerably, but also the server that needs to be analyzed may be on the other side of the world!

I am not saying that the plug is never pulled; rather, the decision is not black and white anymore. Accordingly, there are a few ways to respond to live data collection: focusing on the *collection of the volatile data, collecting volatile data AND log files* (e.g., IDPS, router, firewall), or conducting a *full investigation by collecting every-thing* (a forensic duplication where every bit of data is copied). A duplication (or disk image) is necessary if court proceedings are imminent. The image is stored on an external drive, or you may send it over the network using a network utility such as netcat or cryptcat. Netcat is a networking utility that reads and writes data across a network connection. Cryptcat is Netcat with encryption. Never use the suspect system to do any analysis. Doing so will overwrite evidence.

For example, if the incident merits collecting the volatile data, the investigator may run a script from a USB drive on the suspect machine. The script will run different commands to determine the following: open ports, who is logged on to the system, dates/times, running processes, applications running on certain ports, unauthorized user accounts and privileges, etc. Then, the script output is directed to a file stored on your USB drive. For example, a script called volatile_collection would look like this:

**E : \ > volatile _ collection > volatile _ data _ case101.txt**

*where the output from volatile_collection script would be stored in the file volatile_ dat_case101.txt.*

Network data, routers, firewalls, IPDS logs, and servers may also need to be analyzed (via network logs) for anything suspicious to determine the scope of the incident, who was involved, and a timeline of events. And, finally, the investigator must identify other sources of data that could go beyond a hard drive, such as a USB drive or mobile phone.

All of this digital evidence collected must be preserved to be suitable in court. Digital evidence is very fragile, like footprints in snow or on the sand, because it is easily destroyed or changed. The FBI suggests having a secure location for storage (locked up), having sufficient backup copies (two suggested), having proof that the data have not been altered (hash algorithm), and establishing a *chain of custody,* which is a written log (aka evidence log) to document when media go in or out of storage (Cameron 2011). In order for the data to be admissible, it has to be proven that they have not been tampered with; therefore, you should be able to trace the location of the evidence from the moment it was collected to the moment it appears in court. If there is a time period that is unaccounted for, there is a chance that changes could have been made to the data. One way to preserve evidence is to transfer digital information onto a read-only, nonrewritable CD-ROM and/or uploading the data onto a secure server and hashing (discussed earlier) the file to ensure the data's integrity.

### 5.4.3 ANALYSIS

Following data collection is data analysis. The image of the suspect machine(s) needs to be restored so that the analysis of the evidence can begin. The image should be restored on a clean (wiped) drive that is slightly larger or restored to a clean destination drive that was made by the same manufacturer to ensure that the image will fit. Next, the review process begins by using one of the forensic tools such as EnCase, (forensic toolkit) FTK, P2 by Paraben, and Helix 3. To recover deleted files, the unallocated space of the image needs to be reviewed. The investigator may find whole files or file fragments since files are never removed from the hard drive when the delete feature is utilized. What is deleted is the pointer the operating system used to build the directory tree structure. Once that pointer is gone, the operating system will not be able to find the file—but the investigator can! Keep in mind, however, that when new files are created, a memory spot is chosen, so there is a chance that the file is written over or at least part of the file. Here is why: All data are arranged on a hard drive into *allocation units* called *clusters*. If the data being stored require less storage than a cluster size, the entire cluster is still reserved for that file. The unused space in that cluster is call *slack space*. There is an example in Figure 5.5.

But, deleting that 20K file frees up the space for a new file. If that cluster is chosen, the new file is going to be written on top of the old file. If the new file is smaller than 20K, then part of that 20K file will be retrievable (Figure 5.6).

Evaluating every file on the restored image can be an arduous task; thus, one trick a forensic examiner will use is identifying files with known hashed values. Known file hashes can be files that are received from a manufacturer for popular software applications. Other known hashes can be from movies, cracking tools, music files, and images. The National Software Reference Library (http://www.nsrl.nist.gov) provides values for common software applications. An investigator may be able to reduce the number of files needed to be analyzed by 90% by using the "hashkeeper paradigm," which assumes that similar files produce the same hash value (Mares 2002):

1. Obtain a list of hash values of "known" files.
2. Obtain the hashes of the suspect files.
3. Compare the two hash lists to match the known files or identify the unknown files.
4. Eliminate the "known" files from the search.
5. Identify and review the unknown files.

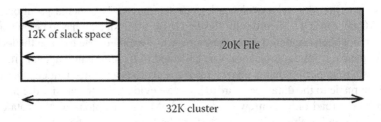

**FIGURE 5.5**   Slack space.

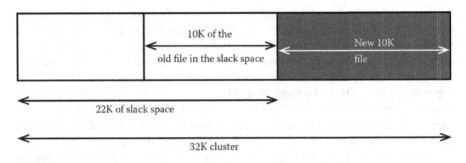

**FIGURE 5.6**   Slack space with a partial file.

For string searches within the data, the drive needs to have all compressed files decompressed and all encrypted files unencrypted. Then, just as you do with web searches, you should use effective key words to pare down results. The exact search methodology used depends on the forensic software tool you are using, what you are the looking for (e.g., files, web browser history, or e-mails), the format of the data, your time constraints, and whether the suspect is aware of the investigation. If so, he or she may have deleted some files.

A common investigation is an Internet usage analysis that monitors inappropriate usage at work where, for example, an employee is gambling or viewing pornography. Divorce lawyers may use this type of analysis to prove infidelity by showing evidence on a social networking site or proving a spouse was on a blog website looking for advice on how to get an easy divorce. An anonymous blog posting can be attributed to a spouse by showing that he or she made purchases with a credit card before and/or after the post on the blog. Web browsers store multiple pieces of information, such as history of pages visited, recently typed URLs, cached versions of previously viewed pages, and favorites. The challenge is also in showing that the accused was actually the one using that computer at the time of the incident. For example, if pornography was viewed on a particular employee's computer, the investigator has to make sure that employee was not on vacation or in meetings all day when the abuse occurred.

When doing this type of analysis (aka a *temporal analysis*), the investigator may want to reconstruct the web page. The web page can be reconstructed by searching the index.dat file for files that are associated with a given URL. Then, the investigator can look for those files in the cache and copy them to a temporary directory. The reconstructed page should be viewed in a browser that is off-line so that the browser does not access the Internet. If the reconstructed page accesses the Internet, it may follow the URL and download the latest version of the page from the server and will not be the version of the web page that the suspect viewed at the time of the incident. Note that the presence of a single image file does not indicate that the individual visited a website. Because there are times when images are a result of a pop-up or redirect, it is important that the investigator also determine which sites were visited prior to the site in question. Recall the case of Julie Amero discussed at the beginning of Chapter 2. These are just a few of the many techniques an investigator could utilize for an effective and forensically sound Internet usage analysis.

E-mail analysis is also very common. The investigator may be tasked to prove a certain policy violation, harassment, or impersonation. It may be as simple as finding the e-mail and determining who sent it and who received it, or as complicated as reconstructing the entire e-mail chain. This is also an analysis that should be done off-line so that no e-mails are sent or received inadvertently during the analysis.

### 5.4.4  REPORTING

The final stage in the DF life cycle is reporting the results of the analysis in the previous stage as well as describing reasoning behind actions, tool choices, and procedures. NIST (Kent et al., 2006) describes three main factors that affect reporting: *alternative explanations, audience consideration*, and *actionable information*.

*Alternative explanations* back up the conclusions of the incident by including all plausible explanations for what happened. In the Julie Amero case, she was convicted (later overturned) of viewing pornography on a school computer in front of minors. Had the investigators looked for alternative explanations, they would have figured out that the pornography websites viewed were caused by spyware. Instead, the clearly ignorant original investigators misled the jury by only presenting evidence in the temporary Internet files directory and the school firewall logs showing that pornography websites were accessed. They missed the alternative explanation and caused her to be unfairly convicted because their findings (temporary Internet files and firewall logs) do not demonstrate a user's intent. Upon reanalysis by expert forensic investigators (Eckelberry et al. 2007), it was discovered that the antivirus software was an out-of-date trial version, that there was no antispyware[3] software installed on the system, and that the spyware was definitely installed prior to the incident.

The original investigators also misled the jury by informing them that spyware is not capable of spawning pop-ups (not true), that pop-ups cannot be in an endless loop (not true), and that the red *link* color used for some of the text on the porn website (that they showed the jury) indicated Amero clicked on the links (the link visits, whether intentional or not, are shown in the visited color, which they indicated was red). In this particular case, the link was red; however, if the original investigators had opened the browser preferences, they would have noted a few things: (1) The links were selected to be green if a site was visited, and (2) the html source code changed the font color to red. The investigators also misled the jury by telling them that the only way spyware is installed on a computer is by actually visiting a pornographic site (not true). Eckelberry et al. (2007) determined that this particular spyware was installed right after a Halloween screen saver was downloaded. There were multiple inconsistencies in the original investigation. I encourage you to read this case as it illustrates very clearly what NOT to do in a forensic investigation.

Reports on the results of a forensics analysis will vary in content and detail based on the incident. Just as in any writing, *audience consideration* is important. Thus, the level of detail of a forensic analysis report is determined by the audience who needs the report. If the analysis resulted in a noncriminal case, the executives or management may want a simple overview of the incident, how it was resolved, and how another occurrence will be prevented. A system administrator may need details regarding network traffic. Or, if it is criminal case, law enforcement will require a very detailed report. In all cases, the report should be accurate, concise, and complete. Nelson, Phillips, and Steuart (2010) suggest that the report should include supporting material, an explanation of the examination and data collection methods, calculations (e.g., MD5 hash), an explanation of any uncertainty and possible errors, and an explanation of the results found and references utilized for the content of the report.

The third report factor is *actionable information*, where the information provided may lead to another source of information and/or information that will help to prevent future similar incidents.

Another type of report is an *after action report* that, in addition to identifying issues that need to be improved upon for future incidents, may also include improvements to the team or process. For example, team members may decide to improve their skills using the forensic software, fix the acceptable-use policy for the organization,

## A GREAT EXAMPLE OF SECURITY THROUGH OBSCURITY[4] (AND DIGITAL FORENSICS)

If you have an e-mail account that you think is anonymous, think again. An anonymous e-mail account is an account that is created without revealing any personal information. However, that alone does not guarantee anonymity. An anonymous e-mail account, when used with mail clients such as Outlook, appends the IP address information to each e-mail's metadata[5] that are sent—which is a great clue for a forensic investigator. Therefore, if the e-mails are investigated, the IP address can help pinpoint the sender. You may be wondering who is looking at your e-mail. Due to the provisions of 1986's Stored Communications Act, the government can access e-mails stored by a third-party service provider. However, there are a few caveats to that access. If the e-mail is in "electronic storage" AND less than 180 days old, a warrant needs to be obtained to access the e-mails. E-mails not in "electronic storage" OR in "electronic storage" more than 180 days can be accessed with a subpoena. Electronic storage is defined as e-mail that has been received by the Internet service provider but has not been opened by the recipient (Jarrett et al. 2009).

The way to make your Internet and e-mail activity really anonymous is to use software such as Tor,[6] which conceals a user's location even if the e-mails are accessed. This type of software is often used by journalists, the military, and activists to protect research, investigations, etc. Hence, Tor needs to be running in order to make an anonymous e-mail account actually anonymous. Unfortunately for a recent CIA director and his girlfriend, this software was not used and their extramarital affair was revealed by piecing together the e-mail trail left by the girlfriend—even though anonymous e-mail accounts were utilized by both of them.

The investigation began due to harassing e-mails sent to another woman (Perez, Gorman, and Barrett 2012). The FBI analyzed the logs from the e-mail provider to determine who sent the harassing e-mails. They specifically looked at the metadata from the e-mails to determine the locations from which the e-mails were sent. By comparing the e-mail metadata from the e-mail providers, guest lists from hotels, and IP login records (hotel WiFi), the FBI put the puzzle together and discovered the identity of the sender of the harassing e-mails as well as her affair with the CIA director (Isikoff and Sullivan 2012). This discovery led to the CIA director's resignation.

or modify the IR procedure. This will, of course, help in staying current with the changes in law, technology, and the latest cyber issues.

## 5.5   MOBILE PHONE FORENSICS

Mobile phone forensics is the science of recovering digital evidence from a mobile phone. Mobile phones, being much more than a communication tool, retain a substantial amount of data: calendars, photos, call logs, text messages, web history, etc. The cellular network also retains data regarding location. Mobile phone data have helped convict many criminals. A man was convicted of the murder of a college student with the help of cellular phone network data (Summers 2003). The killer was actually helping the police locate the college student by showing them around the college campus. At the same time, the police were analyzing the cellular network data. When the victim's phone was turned off, it disengaged itself from a specific cellular tower near the killer's home; due to the mounting evidence, the killer eventually admitted to the crime.

Mobile phone forensics uses the DF life cycle just described. The challenge with mobile phone forensics is the frequent release of new phone models, often making cables and accessories obsolete, as well as the lack of a standard for where mobile phones store messages. The upside is that the puzzle can be more easily solved because of all of the information stored on cell phones: calls (incoming, outgoing, missed), address books, texts, e-mail, chat, Internet use, photos, videos, calendars, music, voice records, notes—the list could go on forever with all the apps available as well.

A brief overview of NIST (Jansen and Ayers, 2007) recommendations to seize mobile phones in an investigation will be presented in this section. For further details, download the special publication listed in the reference section of this chapter. First, consider all of the types of evidence needed from the phone. For example, if fingerprints are needed, follow the appropriate handling procedures for acquiring fingerprints. The investigator may also want to record any viewable information from the phone. It is advisable to leave the phone off for two reasons: First, there is a potential for data loss if the battery dies and, second, data may be overwritten if network activity occurs. If the phone must remain on for some reason, the phone should be placed in a container that blocks radio frequency or in airplane mode. Finally, that container should be placed into a labeled evidence bag that is then sealed to restrict access. Before leaving the scene, collect all related phone hardware such as cradles, cables, manuals, and packaging—anything you find related to the phone.

To collect the evidence from the mobile phone, NIST (Jansen and Ayers, 2007) also recommends isolating the phone from all other devices used for data synchronization. Imaging the device at the scene is the best option if battery depletion is an issue. If not, bring it back to a lab to acquire the data. There are many memory categories as well as various memory structures that vary among manufacturers. Sometimes it is just call log data, so it may not be necessary to recover all of the data on the phone. If you are not familiar with acquiring data from a particular phone, it is best to seek assistance from another digital forensic professional.

## INFRAGARD

A great organization for security professionals is InfraGard—a not-for-profit organization that is a partnership between the private sector and the FBI. The members of InfraGard are individuals from businesses, academic institutions, state and local law enforcement, and any person that wants to participate in sharing information and intelligence that may prevent hostile acts against the United States (http://www.infragard.net). For example, a university professor (and his class) helped the FBI catch criminals involved in a case called "Trident Breach." In this case, the criminals infected computers (via an e-mail link or attachment) with the ZeuS virus, which is essentially a key logger application that logged the user's banking information. Then, the criminals were able to lure other people (aka money mules) into "work-at-home" schemes where their "job" was completing banking transactions.

The banking transaction went as follows: The mule opened the bank account, the money was deposited from the ZeuS-infected computer user's account into the mule's account, and then the mule withdrew the cash and sent it to the criminals. Gary Warner, professor at the University of Alabama at Birmingham (and member of InfraGard), used data-mining techniques to establish the links between the ZeuS-infected computers and the origin of the mass infector. Most of the hackers and the "mules" were caught. The 18 mules still at large were found by his students using computer forensic investigation techniques such as crawling social networking sights to identify the remaining suspects (Engel 2012).

To apply for a membership to InfraGard, fill out the online application (http://www.infragard.net/member.php), read and sign the "Rules of Behavior" form, and submit. Once you are accepted, find your local chapter and attend any meetings of interest.

## REFERENCES

Cameron, S. August 2011. Digital evidence. *FBI Law Enforcement Bulletin.*

Cichonski, P., Millar, T., Grance, T. and Scarfone, K. 2012. Computer security incident handling guide. NIST special publication 800-61, revision 2.

Eckelberry, A., Dardick, G., Folkerts, J., Shipp, A., Sites, E., Stewart, J. and Stuart, R. March 21, 2007. Technical review of the trial testimony. *State of Connecticut v. Julie Amero.* http://www.sunbelt-software.com/ihs/alex/julieamerosummary.pdf (retrieved March 2, 2013).

Engel, R. 2012. University professor helps FBI crack $70 million cybercrime ring. http://rockcenter.nbcnews.com/_news/2012/03/21/10792287-university-professor-helps-fbi-crack-70-million-cybercrime-ring (retrieved August 3, 2012).

FBI.gov. May 31, 2011. Regional labs help solve local crimes. http://www.fbi.gov/news/stories/2011/may/forensics_053111 (retrieved February 9, 2013).

FCC (Federal Communications Commission). 2001. FCC computer security incident response guide.

Isikoff, M. and Sullivan, B. November 12, 2012. Emails on "coming and goings" of Petraeus, other military officials escalated FBI concerns. *NBC News.* http://openchannel.nbcnews.com/_news/2012/11/12/15119872-emails-on-coming-and-goings-of-petraeus-other-military-officials-escalated-fbi-concerns?lite (retrieved March 8, 2013).

Jansen, W. and Ayers, R. 2007. Guidelines on cell phone forensics. NIST special publication 800-101.

Jansen, W. and Grance, T. 2011. Guidelines on security and privacy in public cloud computing. NIST special publication 800-144.

Jarrett, H., Bailie, M., Hagen, E. and Judish, N. 2009. Searching and seizing computers and obtaining electronic evidence in criminal investigations. US Department of Justice. http://www.justice.gov/criminal/cybercrime/docs/ssmanual2009.pdf (retrieved March 12, 2013).

Kent, K., Chevalier, S., Grance, T. and Dang, H. August 2006. Guide to integrating forensic techniques into incident response. NIST special publication 800-86.

Lynch, W. 2005. Writing an incident handling and recovery plan. http://www.net-security.org/article.php?id=775&p=3 (retrieved February 21, 2013).

Mandia, K., Prosise, C. and Pepe, M. 2003. *Incident response & computer forensics*, 2nd ed. New York: McGraw-Hill.

Mares, D. May 2002. Using file hashes to reduce forensic analysis. *SC Magazine.*

Nelson, B., Phillips, A. and Steuart, C. 2010. *Guide to computer forensics and investigations*, 4th ed. Boston: Cengage Learning, Course Technology.

NIST. September 2020. Data integrity: Recovering from ransomware and other destructive events. NIST SP 1800-11B.

Perez, E., Gorman, S. and Barrett, D. November 12, 2012. FBI scrutinized on Petraeus. *Wall Street Journal.*

Summers, C. December 18, 2003. Mobile phones—The new fingerprints. *BBC News.*

Verizon. 2012. 2012 Data breach investigations report. http://www.indefenseofdata.com/data-breach-trends-stats/ (retrieved February 8, 2013).

## NOTES

1 Industrial espionage is an attempt to gain access to trade secrets.

2 Here is an example of disk-wiping software: electronic data disposal: DOD-compliant disk sanitation software (http://www.auburn.edu/oit/it_policies/edd_dod_compliant_apps.php).

3 Antispyware is software designed to detect and remove a malicious application from a computer.

4 Security through obscurity is a derogatory term that implies that secrecy or hiding something makes it secure. It is similar to when I put my laptop on the front seat of my car and cover it by a blanket while parked in a public parking lot. If it is discovered, there is nothing really protecting it from getting stolen.

5 Metadata are data that describe other data.

6 Tor (https://www.torproject.org/) is software that was developed originally to protect government communication. Now it is used by people as a safeguard to their privacy (not anonymity). It is used to safeguard a person's behavior and interests. One of the examples on the Tor website is when traveling abroad, Tor can hide your connection to your employer's computer so that your national origin is not revealed.

# 6 Development, Security, and Operations

A good programmer is someone who always looks both ways before crossing a one-way street.

*—Doug Linder*

## 6.1 WHAT IS A SECURE SOFTWARE DEVELOPMENT LIFE CYCLE?

The software development life cycle (SDLC), sometimes known as the secure software development life cycle, is an effective way to focus on the protection of information and information systems—especially when the integration of security is a focus in every step of the software system development process. The SDLC is a multistep, iterative process (shown in Figure 6.1), starting with the analysis, followed by the design, and implementation steps, and continues through the maintenance and disposal of the system (Radack 2009).

The SDLC describes the process for building information systems in a methodical way during every stage of a products life. According to Elliott and Strachan (2004), the SDLC "originated in the 1960s, to develop large-scale functional business systems in an age of large-scale business conglomerates. Information systems activities revolved around heavy data processing and number crunching routines."

Other system development frameworks have used the SDLC as a foundation, such as the structured systems analysis and design method produced for the UK government Office of Government Commerce in the 1980s. Ever since, according to Elliott and Strachan (2004), "the traditional life cycle approaches to systems development have been increasingly replaced with alternative approaches and frameworks, which attempted to overcome some of the inherent deficiencies of the traditional SDLC."

## 6.2 REASONS TO USE SDLC

The goals of an SDLC approach are the following (SDLC Forms 2021):

1. Deliver quality systems that meet or exceed customer expectations when promised and within cost estimates.
2. Provide a framework for developing quality systems using an identifiable, measurable, and repeatable process.
3. Identify and assign the roles and responsibilities of all involved parties, including functional and technical managers, throughout the system development life cycle.

DOI: 10.1201/9781003245223-6

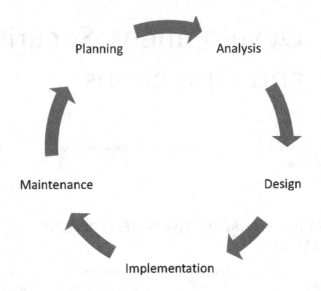

**FIGURE 6.1** The software development life cycle.

4. Ensure that system development requirements are well defined and subsequently satisfied.
5. Establish appropriate levels of management authority to provide timely direction, coordination, control, review, and approval of the system development project.
6. Ensure project management accountability.
7. Document requirements and maintain traceability of those requirements throughout the development and implementation process.
8. Ensure that projects are developed within the current and planned information technology infrastructure.
9. Identify project risks and issues early and manage them before they become problems.

## 6.3 SEGREGATION OF ENVIRONMENTS

Although secure development operations can be about removing barriers between systems, segregation is about adding barriers to minimize risk. Thus, segregating IT environments is a technique that is still relevant today. Segregation serves as an obstacle to bad actors as it makes lateral movement difficult and isolates security issues. Whether it be a malicious attack or technical fault, proper segregation of IT environments can limit the spread to other internal areas, reducing the potential impact (Bickerstaffe 2022). In addition, development should be performed in a dedicated network zone and should be separate from quality assurance and production. In addition, to being best security practice, if you are subject to most regulatory requirements or undergo audits of the software, a finding will be issued if you fail to follow this practice.

## 6.4 SECURE SDLC PHASES

There can be multiple descriptions of phases in various methodologies (shown in Figure 6.2), but essentially, they follow these basic tasks:

*Initiation or Planning:*
    During this phase, identify and validate an opportunity to improve business accomplishments or deficiencies related to a business need. Significant assumptions and constraints on solutions will be identified and recommendations to explore alternative concepts and methods to satisfy the need will be given.
*Feasibility:*
    The feasibility phase is the initial investigation or brief study of the problem to determine whether the systems project should be pursued. A feasibility study establishes the context through which the project addresses the requirements and investigates the practicality of a proposed solution. The feasibility study is used to determine if the project should get the go-ahead.

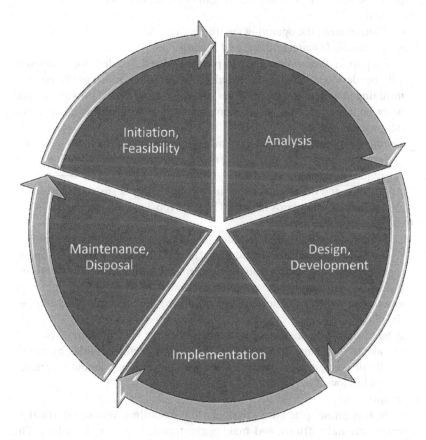

**FIGURE 6.2** Phases.

If the project is to proceed, the feasibility study will produce a project plan and budget estimates for the future stages of development.

*Requirements Analysis:*

This phase formally defines the detailed functional user requirements, using high-level requirements identified in the *initiation* and *feasibility* phases. In this phase, the requirements are defined to a sufficient level of detail for systems design to proceed. Requirements need to be measurable, testable, and relate to the business need or opportunity identified in the initiation phase.

*Design and Development:*

During this phase, the system is designed to satisfy the functional requirements identified in the previous phase. Since problems in the design phase can be very expensive to solve in later stages of the software development, a variety of elements are considered in the design to mitigate risk. These include:

a.  Identifying potential risks and defining mitigating design features;
b.  Performing a security risk assessment;
c.  Developing a conversion plan to migrate current data to the new system; and
d.  Determining the operating environment.

*Implementation, Documentation, and Testing:*

As part of the implementation phase, updated detailed documentation will be developed and will include all operations information needed, including detailed instructions for when systems fail. Software should never be moved into the production environment without this *documented* information.

*Testing* should be performed in its own environment and includes unit, integration, and system testing to endure the proper implementation of the requirements

Application and infrastructure security vulnerability scans should be based on industry standards (such as OWASP Top Ten vulnerabilities).

*Operations and Maintenance:*

System operations and maintenance should be ongoing. An annual review with stakeholders should be performed. Systems should be s monitored for continued performance in accordance with user requirements and needed system modifications are incorporated when identified, approved, and tested. When modifications are identified, the system should re-enter the planning phase.

A team should be designated to manage *security vulnerabilities*, who will identify, manage, and minimize the security vulnerabilities by a code fix or configuration change. A security patch policy should be created and followed.

*Disposal:*

In this phase, plans are developed for discarding system information, hardware, and software and making the transition to a new system. The information, hardware, and software may be moved to another system, archived, discarded, or destroyed. If performed improperly, the disposal phase can result in the unauthorized disclosure of sensitive data. When archiving information, organizations should consider the need for and the methods for future retrieval (Radack 2009).

## 6.5 WHY DO DEVELOPERS NOT FOLLOW SDLC?

The SDLC as shown above has very clear steps, processes, and procedures to ensure a secure and successful software deployment, yet as can be seen by the ongoing list of breaches (Black Kite Research 2022) software vulnerability is still a major problem. Black Kite, creators of the cybersecurity risk rating system (2022), believes it is a mentality of "seems good enough" is the root cause. As developers, you must not fall prey to that thought process, you must follow the SDLC and bakes security at every phase.

## 6.6 IS SDLC AN OODA LOOP?

Is the SDLC the same as the OODA Loop? If you compare them side by side (Figure 6.3), they do look similar, but there are some very clear characteristics that make them very, very different. In Figure 6.3, one can see there are four major sections: (1) observe, (2) orient, (3) decide, and (4) act. But an OODA Loop is far more than simply going from one step to another. For instance, in the *orient* stage, you have many factors that feed into that orientation, such as cultural traditions, genetic heritage, analysis and synthesis, new information, and previous experience. When we look at the SDLC graphic we see very clearly that there are phases that lead one into the other. From a very high level, yes, they are similar, even though SDLC does seem to be linear without the potential loop interruptions such as Implicit Guidance and Control, Feed Forward, Feed Back, and Unfolding interaction with the environment.

Each phase of the SDLC is essentially in and of itself an OODA Loop. Let us look at the analysis phase. In part of this phase, you formally define the detailed functional user requirements. In OODA terminology you will *orient* to the data you

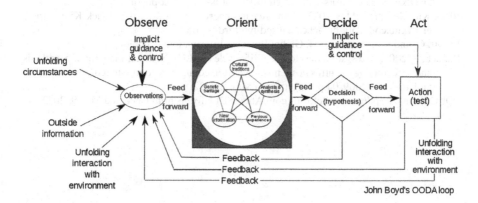

**FIGURE 6.3** SDLC/OODA comparison.

collected in the *observe* stage. During *observe*, high lever requirements are collected. Those high-level requirements may be from multiple users and stakeholders with different needs from the same application. In the orient phase, measurable requirements need to be created. They also need to be testable and relate to the business need or opportunity identified previously. You may discover an analogous project that comprises requirements that have been collected from a previous project that fit the current project—thus you can move to the next phase without further analysis. The OODA Loop term for this is *Implicit Guidance and Control*, in other words previously learned knowledge that allows you to advance the loop faster.

> You may also discover that not enough information was observed to enable a quality result in orient phase of the OODA Loop, which would again promote engaging in another form of Implicit Guidance and Control and revert to the observation phase to collect additional information.
>
> At some point, you will decide that the requirements are measurable, testable, and related to the business need and feed forward to a decision. That decision, to move into the design phase is the **act** in the OODA Loop. OODA's unfolding *interaction with environment* is simply the concept that now we have been moved into the design phase, a flaw in the requirements may surface and require a return to the requirements phase and start new observations.
>
> The SDLC provides a structured framework to follow to endure the development of security-focused code while practicing the decision-making processes in the OODA Loop, and it will allow products on a faster road to market.

## REFERENCES

Bickerstaffe, E. 2022. Security Think Tank: Proper segregation is more important than ever. https://www.computerweekly.com/opinion/Security-Think-Tank-Proper-segregation-is-more-important-than-ever.

Black kite. 2022. Controls without enforcement: Is zero trust POSSIBLE? Black Kite: https://blackkite.com/controls-without-enforcement-is-zero-trust-possible/.

Black Kite Research. May 2022. Data breaches caused by third-parties. Black Kite: https://blackkite.com/data-breaches-caused-by-third-parties/.

Elliott, G. and Strachan, J. 2004. Global Business Information technology, p. 87.

Radack, S. 2009. The system development life cycle (SDLC). National Institute of Standards and Technology. https://csrc.nist.gov/csrc/media/publications/shared/documents/itl-bulletin/itlbul2009-04.pdf (Retrieved May 7, 2022).

SDLC Forms. http://www.sdlcforms.com/AboutSDLCforms.html (accessed May 9, 2022).

# 7 Mobile Device Forensic Tools[1]

Technology and tools are useful and powerful when they are your servant and not your master.

—*Stephen Covey*

## 7.1 INTRODUCTION

The intent of this chapter is to introduce some vendor-based and free tools available for mobile forensic device analysis and discuss the basic tasks performed in a mobile device forensic investigation. The tools discussed in this chapter are not in any order or preference as there are many options available to an investigator or analyst. In addition, the tools discussed in this chapter are not inclusive or representative of all that is available to perform the tasks necessary for an effective mobile device forensic investigation.

Unlike computer forensic analysis tools, there are very few freeware or open-source tools capable of decoding the volumes of potentially relevant data that can be associated with a mobile device. One of the main reasons for this is that mobile device applications and application-based data are the proverbial "wild card" when it comes to mobile device acquisition and analysis. Currently, there are over 3,000,000 applications on the Google Play Store and over 2,000,000 applications on the Apple App store. These numbers will only increase over time. Programming, reverse-engineering, and decoding all of these applications is a monumental task that even the largest cellular device forensic software companies do not have the manpower, time or resources to facilitate. Additionally, applications in use on mobile devices are updated regularly, and the changes that are sometimes part of these updates would require a vast overhaul of how the mobile forensic tool decodes the data within the application. Because of these factors, mobile forensic software companies generally concentrate their decoding efforts on the most popular applications, a definition and direction that also changes over time. This also means that the analyst will likely have to conduct manual analysis of certain applications at some point during an investigation and will accordingly have the skills to acquire, recognize, and analyze data from within the application SQLite databases where the vast majority of application data is stored.

Another significant hurdle that can be part of the mobile forensic process is data acquisition. Unlike computer forensics, where the acquisition of data is generally more straightforward, mobile devices—and particularly Android devices—operate on a variety of hardware manufacturer platforms. This can lead to challenges when attempting to acquire data. Furthermore, the advent of heightened data security measures across the mobile device spectrum has led to increased difficulty when attempting to acquire mobile data in certain circumstances. Increasingly, the data that is of use to us as analysts is stored in a protected area of the phone memory, which is mostly inaccessible without the user pass code and/or removing certain security measures, which can lead to questions about evidence handling later in the process.

The following is a current listing of considerations when attempting to acquire mobile device data:

- Prior to acquisition, ensure any and all physical evidence, such as finger-prints, DNA, and blood have been collected from the item(s). Consider the use of sterile gloves during the handling of the device.
- Obtaining the pass code to the device is vital in most circumstances. For both Android and Apple mobile devices, access to the device and bypassing certain security measures are necessary for the successful acquisition of the data. This is generally only obtained via the passcode to the device.
- All possible data acquisition methods should be attempted. This can be dictated by the capabilities of the acquisition tool you're using (summarized in Figure 7.1), but can include:
  - Physical acquisition, including micro-read, chip-off, and binary/hex dump
  - Logical/Advanced Logical acquisition
  - File system acquisition
  - Manual Extraction (photographing the phone screen and scroll)
  - SIM card module acquisition
  - Micro SD (expandable memory) card acquisition
- In most circumstances, powering off the device is not advisable. Ensure the device is disconnected from any network connection (including Bluetooth and wireless internet). This can be accomplished by placing the device into airplane mode and verifying no connection and/or placing the device in a signal-blocking package.
- Advanced extraction methods through tools such as Gray Key, Cellebrite Premium, and Belkasoft may be possible (discussed later), depending on the model of phone and software version. These advanced extraction methods often provide a richer data set for analysis.

**FIGURE 7.1**   The mobile forensic data extraction triangle.

## 7.2 TOOLS

### 7.2.1 Axiom and Axiom Cyber by Magnet Forensics

*Download Link:* https://support.magnetforensics.com/s/software-and-downloads

Axiom is a universal tool for computer, cloud, and mobile forensic analysis. It has the ability to ingest and process data from multiple sources, operating systems, and user platforms. For mobile forensics specifically, it can be paired with the freeware Magnet Acquire tool to acquire the data from the mobile device or the acquisition can be conducted using another method and the data ingested into Axiom for processing and analysis (illustrated in Figure 7.2). Axiom is capable of natively decoding a large number of third-party applications on both Android and Apple devices, as well as manual analysis of the databases that store data not natively decoded.

Accompanying Axiom/Axiom Cyber is a utility for data acquisition. Magnet Forensics also offers freeware for customers for data acquisition in the stand-alone utility called Axiom Acquire, which can not only acquire data from mobile devices (iOS and Android) but also from computer and peripheral devices as well.

Additionally, Axiom can ingest all of the computer data (PC, Mac, and Linux) into a case along with the mobile data and present it in a very easily sorted and readable format. Axiom also has a feature called "Portable Case" which allows the analyst to export relevant case details (or a dump of the case data) into a freeware version of Axiom for third-party review, lightweight analysis, and report generation.

Axiom is equipped with two internal applications: Axiom Process and Axiom Examine. Axiom process can acquire, identify, and process the evidence files in either default or user-defined categories, such as applications, document types, and settings, on either mobile or computer devices (illustrated in Figure 7.3). The Axiom Examine utility is the main examination tool used by the analyst to conduct keyword analysis, application analysis, file system analysis, and generate reports for review by stakeholders in the case.

Because there is much crossover between mobile and cloud data, Axiom also incorporates the ability to collect and analyze cloud-based data from any number of sources. The biggest warning with access to cloud-based data is obtaining proper legal authority. The analyst must have proper legal authority in order to access the cloud accounts. This is usually accomplished by search warrant, court order or owner consent. Beyond cloud data, Axiom can also process and help the analyst review

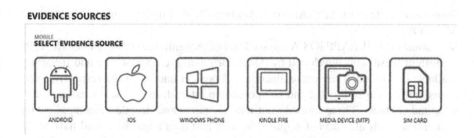

**FIGURE 7.2** Mobile evidence source options in Axiom Cyber.

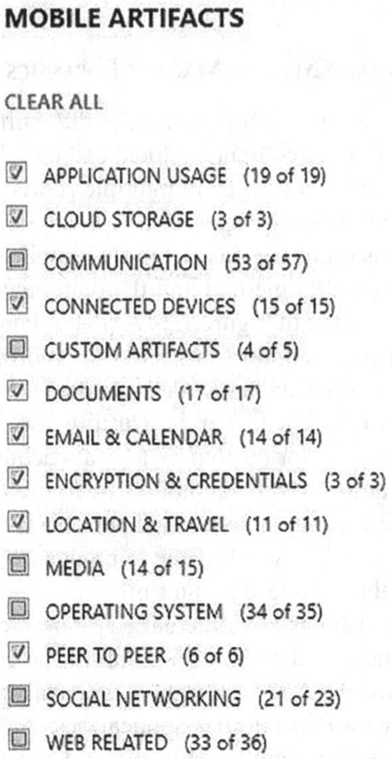

**MOBILE ARTIFACTS**

CLEAR ALL

☑ APPLICATION USAGE  (19 of 19)

☑ CLOUD STORAGE  (3 of 3)

☐ COMMUNICATION  (53 of 57)

☑ CONNECTED DEVICES  (15 of 15)

☐ CUSTOM ARTIFACTS  (4 of 5)

☑ DOCUMENTS  (17 of 17)

☑ EMAIL & CALENDAR  (14 of 14)

☑ ENCRYPTION & CREDENTIALS  (3 of 3)

☑ LOCATION & TRAVEL  (11 of 11)

☐ MEDIA  (14 of 15)

☐ OPERATING SYSTEM  (34 of 35)

☑ PEER TO PEER  (6 of 6)

☐ SOCIAL NETWORKING  (21 of 23)

☐ WEB RELATED  (33 of 36)

**FIGURE 7.3**   Available mobile artifacts in magnet Axiom Cyber.

vehicle-based data from tools such as Berla, which collects data from vehicle info-tainment systems.

As manual analysis of mobile applications is frequently an issue, Axiom/Axiom Cyber also has built-in capability for analysis of app-based SQLite database tables for manual review and analysis. Axiom is in use by many law enforcement agencies and private sector entities across the globe for computer and mobile forensic analysis.

### 7.2.2   ALEAPP ANDROID AND IOS ANALYSIS TOOLS

*Download Link:* ALEAPP Android Analysis Tool: https://github.com/abrignoni/ALEAPP

*Download Link:* iLEAPP iOS Analysis Tool: https://github.com/abrignoni/iLEAPP

Both of these tools developed by Alexis Brignoni are open-source and freeware, available on Github. They parse, decode and present data from the respective plat-forms in the areas of highest and most common interest, such as text messages, call logs, and contacts. These tools do not natively parse and decode much application data. These tools also do not acquire the data and data acquisition will have to be performed by another tool. ALEAPP requires Python scripting tool to be installed. iLEAPP also can require Python for use, but an executable program can be compiled following the steps on Brignoni's Github portal. As encryption is routinely in place with both Apple and Android devices, it should be noted that neither of these applications has the ability to natively decrypt datasets, even with the password. Regardless, they are a decent option for decrypted analysis of both iOS and Android datasets for a high-level view or preview.

### 7.2.3 BELKASOFT EVIDENCE CENTER X

*Download Link:* https://belkasoft.com/get

Belkasoft is a paid commercial tool that focuses it's engineering and development on Apple iOS platforms; however, it can also be used for Android OS analysis. Belkasoft offers a full file system extraction of richer data set for a variety of Apple iOS devices and is used by private sector and law enforcement agencies all over the world. Evidence Center X also has the capability to acquire and analyze data from computer and cloud-based sources, as many of the tools now do. Frequently, the developers of Belkasoft are on the initial stages of testing, updating, and release of the tool, which makes this option an appealing one for analysts, as operating systems, hardware, and application versions are constantly changing.

### 7.2.4 CELLEBRITE UNIVERSAL FORENSIC EXTRACTION DEVICE (UFED) AND PHYSICAL ANALYZER

*Company website:* https://cellebrite.com/en/home

Cellebrite offers it's mobile forensic data acquisition and analysis tools to both government and private sector analysts in a number of different formats. Currently, the options for data acquisition are Cellebrite's UFED Touch 2 (stand-alone unit) and UFED for PC (computer-based unit). The main differences between these two options are that the UFED Touch 2 is a compact, portable system that can be deployed easily in remote or on-site applications, which is ideal for law enforcement on the site of a search warrant or in a larger enterprise environment. The UFED for PC is more ideal for laboratory use because it leverages the power of the user's computer system, which can vary greatly, depending on the options the user has built into their computer system. While the UFED for PC can be deployed on a laptop computer for field use, the UFED Touch 2 is still considered more compact, portable, and a better option for field use. Each kit also comes with a library of connectors and cables to accommodate any and all current and legacy devices, which is another benefit to Cellebrite's offerings. While the modern era has seen a great decrease in the number and type of phone connectors, chargers, and physical cable interface types, the ability to acquire data using UFED Touch 2 or UFED for PC using this universal kit of connectors is a vital one, when encountered either in the lab or the field. Either tool allows for varying levels of acquisition, from physical to advanced logical to manual (picture grab). They also allow for acquisition of SIM cards and extra storage devices, such as Micro SD cards. The equipment for these acquisitions is also included in the Cellebrite UFED Touch 2 or UFED for PC kit.

Once the data is acquired through Cellebrite UFED Touch 2 or UFED for PC, the main tool for analysis of the data garnered is Cellebrite Physical Analyzer (PA), which allows for a deep-dive into the mobile data, as well as a higher-level view. In Cellebrite PA, there are natively parsed and decoded areas known as "Analyzed Data," which offers a fully decoded view of standard messages, application-based messages, pictures, videos, web history, contact information, and call logs, among other items (illustrated in Figure 7.4). The analyst is also able to view the installed applications and sort which are natively decoded by Cellebrite and which are not.

One newer feature to PA is the "App Genie," which uses artificial intelligence to help decode non-native applications. Along with that, there is an SQLite database wizard, which assists in manual analysis of both natively and non-natively decoded applications. Physical Analyzer also has the ability to capture and utilize keys for access to cloud-based accounts, including social media data. These keys can then be utilized within Physical Analyzer to acquire the cloud-based user data, but just as with Axiom, proper legal authority for this collection is of paramount importance.

Much like Axiom, Cellebrite also has a version of a freeware portable case review tool called Cellebrite Reader. A "report" can be generated in Cellebrite Reader and exported to share with other stakeholders in the case. This data can be pre-sorted to case-specific parameters in Physical Analyzer prior to generating the Cellebrite Reader report or it can simply encompass a "data dump" of all of the available data on the device. The Cellebrite Reader application then accompanies the data generated so a user who does not have a full license for Cellebrite can review the data within the Reader program.

Other benefits of Cellebrite licensure are the ability to scan for malware within the data acquired and the access and use of Cellebrite Premium, which is a per-device service offered for advanced data extraction and pass code bypass extraction.

### 7.2.5 OXYGEN FORENSICS

*Download Link:* https://www.oxygen-forensic.com/en/order

Virginia-based Oxygen Forensics is another strong offering in mobile forensic data acquisition and analysis. As mobile forensic tool developers go, they offer their competing suite of tools including Oxygen Detective, Cloud Extractor, Kit, and associated Enterprise versions of their offerings. The difference with Oxygen would be that they offer the individual packages a-la-carte, so the analyst can choose the options they would like to add to their toolbox or as a group, which is often much less expensive than other tools on the market. When faced with mounting costs of these tools, the ability to choose which options to incorporate into the laboratory and flexibility with cost and space, the options that Oxygen offers are appealing.

Oxygen also focuses some development on application-specific offerings for WhatsApp and has developed an exploit for Apple iOS full file system extraction and drone support, as many of the other tools mentioned here have done. With many of the same capabilities as higher-priced alternatives such as SQLite database analysis, cloud data extraction (with or without keys), and portable case review capability, Oxygen presents a very good option for those who are looking for value in their mobile forensic analysis toolbox.

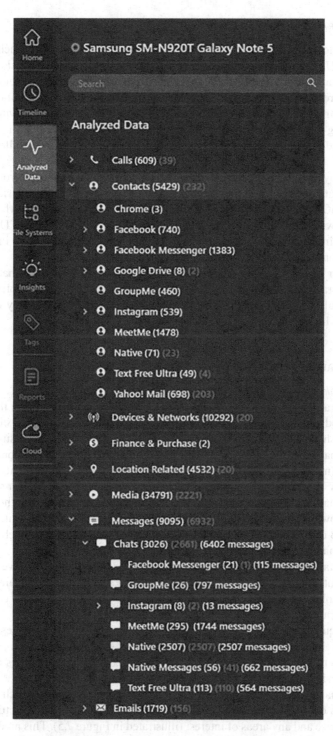

**FIGURE 7.4** Example of analyzed data in Cellebrite Physical Analyzer.

## 7.2.6 Graykey from Gray Shift

Graykey (note: no download link available) is a stand-alone unit, much like the Cellebrite UFED Touch 2, which offers passcode bypass and data extraction for both Apple iOS and Android devices. Graykey is only available to law enforcement, military, and prosecutor's offices in a strict license agreement with these entities. Unlike all of the other tools mentioned in this chapter, Graykey does not assist with analysis or decoding—the main function of the tool is for data extraction only, which can then be imported into your mobile forensic tool of choice, such as Axiom, Cellebrite, and Oxygen.

The Graykey device is under strict lockdown and cannot be moved from it's registered location. Under certain circumstances, the device will obtain a rich dataset from both iOS and Android devices in a full file system data extraction, which can be a dataset twice the size or more of an advanced logical data extraction. This means it is acquiring much more data to be analyzed, which is often important in criminal investigations.

Whether or not you have access to Graykey, the possibility exists that you will need to extract and analyze a Graykey extraction, therefore choosing a software platform that allows for appropriate decoding and analysis of the Graykey data is of paramount importance.

## 7.2.7 DataPilot from Susteen

*Vendor Link*: https://datapilot.com/

The DataPilot operates on a Windows IOT handheld device platform that is ruggedized and suitable for use in the field. The device and accessories are stored in its own hard-shell case for shipping and/or transport. Connection to the desired mobile device is made directly through one of a set of cables that are supplied along with the DataPilot (micro USB, USB-C, Apple lightning, and Apple 30-pin). The DataPilot also comes with a docking station, which can be connected to a computer system via USB for charging, and an extra battery which is easily exchanged with the installed battery. However, the DataPilot can also be charged as a stand-alone unit, using the micro USB connection. The system is also equipped with a camera, which is used to facilitate manual (camera-based) acquisition.

The DataPilot offers several options for logical data extraction to give the highlights of potential interest of the data on the device. Encryption of acquired data is also available, which can be crucial when dealing with certain types of data on the iOS platform.

Fast acquisition method gets the basics—contacts, calls, and messages, which is valuable if time is short. The "Flex Acquisition" offers check-box options of categories (contacts, call history, files, etc.) from which to cherry-pick in cases of limited legal authority, time constraints, etc.

The final type of available acquisition is Optical Capture, which uses the DataPilot's on-board camera to compile a report of a series of screen pictures of the mobile device and any areas of interest (illustrated in Figure 7.5). This mode method would also be useful for capturing cloud-based data that is not available via data

**FIGURE 7.5** Optical capture of iPhone with DataPilot from Susteen.

extraction, for validating the existence of apps on the device and/or the hardware information and for documenting apps that are not supported for analysis by the DataPilot.

The DataPilot seeks to solve the problems of time and backlog by streamlining the mobile forensic data acquisition process and give the analyst a basic view of the data on the device. The DataPilot is not a laboratory tool. It is a field tool that makes acquisition simple and effective for those who find themselves in an environment where they may not have access to the device for an extended period and just want to grab the data and go.

### 7.2.8 XRY FROM MICRO SYSTEMATION

*Vendor link:* https://www.msab.com/products/

The advent of mobile forensics as a practice combined with the ubiquitous nature of devices and their involvement in virtually every type of investigation has caused mobile forensic software tool companies like Swiss-based MSAB to create the proverbial "easy-button" to find the most often used and relevant data more quickly. Whatever term the specific forensic tool uses for this area of data analysis, it generally allows browsing data and gives the 10,000-foot view of the common areas of evidence such as calls, contacts, and text messages. XRY (illustrated in Figure 7.6) color-codes these areas in an easy-to-navigate menu which also displays the overall number of these items (so you know how much time you might spend viewing them).

Also available in XRY are "Quick Views" where we can filter by file or data type, date, deleted items, and more (illustrated in Figure 7.7). This is a handy feature to help streamline your investigations and save time.

**FIGURE 7.6**   Evidence summary in XRY.

One slightly advanced area of analysis that XRY also handles well is the ability to locate, browse, and analyze SQLite databases. By navigating to the "Databases" area, we can browse which databases are part of our extraction data and conduct a search to locate which ones we may want to view. Once located, the list of tables in the database is presented to us and is easily navigable to not only analyze but help validate what the easily parsed information may be showing us (illustrated in Figure 7.8).

Because SQLite database analysis is so important in mobile forensic device analysis, it is very helpful to have a tool that will not only acquire the data but allow the examiner/analyst to locate the desired database files and analyze them within the tool graphical user interface. All of these may be done independently and with freeware tools but having them in one place cuts down on time and device backlog.

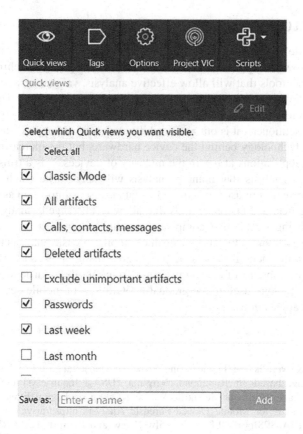

**FIGURE 7.7** Selected artifact sorting in XRY.

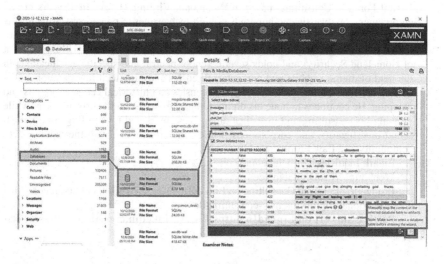

**FIGURE 7.8** Analysis of SQLite databases in XRY.

## 7.3   CONCLUSION

In most mobile forensic analysis, we are concentrating on three main areas of importance—ownership, activity, and location. Accordingly, researching, choosing, and investing in tools that will allow effective analysis, validation, and reporting of these items helps the analyst choose which tool(s) will best suit their needs.

Mobile forensic data acquisition and analysis has turned into a very competitive market. As practitioners, it is our responsibility to know and appreciate not only the ever-evolving technology behind the device hardware, but the increasingly complex nature of the applications at use in the majority of devices. It is a truism of mobile device forensic analysis that manual analysis will need to be conducted at some point. The choice of tool used to acquire the data is just as important as the tool used to analyze the data and choosing tools that allow a full range of analysis options is important to being an effective, competent, thorough analyst.

As a final disclaimer, it is also a truism of mobile forensic analysis that the hardware and software, including applications, are ever-changing. As such, the competent analyst will keep abreast of current offerings from mobile forensic tool vendors, as well as the technology used in the graduating iterations of mobile device hardware, software, and applications.

## NOTE

1 Patrick Siewert is the Founder and Principal Consultant of Pro Digital Forensic Consulting, based in Richmond, Virginia (USA). Inquiries@ProDigital4n6.com. Excerpts taken from the Scientific Working Group on Digital Evidence (SWGDE) Best Practices for Mobile Device Forensic Analysis https://drive.google.com/file/d/ 1lkj3lRnIZAu8PSIp0rFUL8KCViLA5IwO/view and Siewert, P., "XRY v. 9.3 from MSAB" Forensic Focus, https://www.forensicfocus.com/reviews/xry-v9-3-from-msab/, January 18, 2021.

Chapter Author: Patrick Siewert is the Founder and Principal Consultant of Pro Digital Forensic Consulting, based in Richmond, Virginia (USA). In 15 years of law enforcement, he investigated hundreds of high-tech crimes to precedent-setting results and continues to support litigation cases and corporations in his digital forensic practice. Patrick is a graduate of SCERS (US Secret Service) and BCERT (US Department of Homeland Security) and holds several vendor-neutral and specific certifications in the field of digital (mobile and computer) forensics and high-tech investigation and is a court-certified expert witness in digital forensics and historical cell site analysis and mapping. He continues to hone his digital forensic expertise in the private sector while growing his consulting and investigation business marketed toward litigators, professional investigators, and corporations, while keeping in touch with the public safety community as a Law Enforcement Instructor.

Email: Inquiries@ProDigital4n6.com

Web: https://ProDigital4n6.com

Pro Digital Forensic Consulting on LinkedIn: https://www.linkedin.com/company/ professional-digital-forensic-consulting-llc

Patrick Siewert on LinkedIn: https://www.linkedin.com/in/patrick-siewert-92513445/

# 8 The Laws Most Likely to Affect IT and IT Security[1]

> It takes 20 years to build a reputation and five minutes to ruin it. If you think about that you'll do things differently.
>
> —*Warren Buffett*

## 8.1 INTRODUCTION

This chapter will provide a summary of certain laws that have some IT requirements or impact. It is NOT a substitute for legal advice, which only a licensed attorney can provide. Instead, the focus will be on the general purpose of the law with special emphasis on those elements of particular interest to chief information security officers (CISOs).

## 8.2 MANAGING PERSONAL DATA

As daily headlines remind you, many laws relate to the security of personal data. Perhaps the best known are the state data breach laws, as they were the first to require notification to the affected consumers.

### 8.2.1 DATA BREACH LAWS

California enacted the first data breach notification law over 20 years ago and over the next 20 years, every state enacted a similar law. Unfortunately, no two laws are identical, so what constitutes a "breach of the security of the system" in one state may not be a breach in another state.[2] The laws typically do not require any specific security standards, but usually require that the consumer notification explain how the breach occurred. The laws usually require that there be unauthorized "access to" or "acquisition of" personal data. Although "personal data" was originally narrowly defined—usually full name plus Social Security Number, driver's license number, or credit card and security code or financial information enabling access to an account—over the years, legislatures have expanded the term to include date of birth, social media credentials, mother's maiden name, health information, biometrics, and others.

*Relevance to IT*: If there is a successful attack on your systems, one of the first questions you will be asked is whether personal data was affected. If possible, map the locations of personal data in your systems now, so that you can answer the question quickly if an incident occurs. In addition, if you are not required to retain data— as advised by your attorneys—delete it, so that the data is not waiting for a threat

DOI: 10.1201/9781003245223-8

actor to steal. Finally, if there is a successful attack, you will need a good under-standing of what happened, in order to explain it to the regulators. In-house IT staff typically are not trained in forensic analysis and may inadvertently destroy relevant evidence. Instead, if you can, isolate the affected systems and—working with your attorneys—call in the forensic experts.

### 8.2.2 CYBERSECURITY PROTECTION: MASSACHUSETTS/NY SHIELD ACT

A few states have enacted IT security/cybersecurity standards. Massachusetts enacted its standards as part of its breach law, but included many specific requirements not only for protecting personal information, but also for computer system security. The Massachusetts requirements apply to companies that have personal information of Massachusetts residents, regardless of whether the company itself is located in Massachusetts. Since 2010, the regulation requires, among many other things:

> A comprehensive information security program that is written in one or more readily accessible parts and contains administrative, technical, and physical safeguards that are appropriate to (a) the size, scope and type of business of the person obligated to safeguard the personal information under such comprehensive information security program; (b) the amount of resources available to such person; (c) the amount of stored data; and (d) the need for security and confidentiality of both consumer and employee information.[3]

The computer system requirements range from access limitations to encryption to passwords, and many other topics. Massachusetts requires an annual update to the program.

Similarly, in 2019, New York amended its breach law and also enacted security requirements in a law known as the SHIELD Act, which is an acronym for Stop Hacks and Improve Electronic Data Security Act.[4] The law requires that any com-pany that has personal information of even one New York resident to implement reasonable administrative, physical, and technical safeguards in "to protect the security, confidentiality and integrity of the private information including, but not limited to, disposal of data." The law requires covered companies to conduct risk assessments, as well as protect against unauthorized access. The law does not permit private individuals to file a lawsuit but enables the Attorney General to do so. In January of 2022, the New York Attorney General entered into a $600,000 settle-ment vision-benefits-provider EyeMed Vision Care, Inc.[5] because EyeMed (1) had not implemented multi-factor authentication for e-mail; (2) had required 8-character passwords on the e-mail account that the threat actor attacked, when the company had implemented 12-character requirements on other accounts; (3) did not retain logs longer than 90 days; and (4) retained personal information in the hacked e-mail account for 6 years.

*Relevance to IT:* Ask your lawyer which requirements apply to your organization. The laws tend to be very general, because the requirements and expectations for a multi-national company will be very different than for "two guys in a garage" start-ing a tech company.[6] Together, you can determine what the appropriate requirements should be. This could also be a good time to revamp the record retention policy, as

New York just provided 600,000 reasons why retaining personal data that is not necessary is a very expensive problem.

### 8.2.3   CCPA AND CPRA

If your organization is subject to the California Consumer Privacy Act (CCPA), you are already probably familiar with CCPA's requirements that consumers have the right to know the information personal information your organization has on that consumer, and to request deletion of some of that data, including that information held by your organization's service providers (subject to many exceptions). You may also be familiar with the "Do Not Sell My Personal Information" link so that consumers can require the organization not "sell" their personal information to third parties (CCPA defines a "sale" very broadly—where your organization receives something of value in exchange for the consumers' personal information). CCPA does not contain any specific IT requirements, but it does permit private individuals to file a lawsuit under CCPA if the personal information

> is subject to an unauthorized access and exfiltration, theft, or disclosure as a result of the business's violation of the duty to implement and maintain reasonable security procedures and practices appropriate to the nature of the information to protect the personal information.[7]

The law permits the affected consumers to recover the greater of (1) actual damages or (2) an amount between $100 and $750 per consumer per incident, depending upon a variety of factors, including "the nature and seriousness of the misconduct, the number of violations, the persistence of the misconduct, the length of time over which the misconduct occurred, the willfulness of the defendant's misconduct, and the defendant's assets, liabilities, and net worth."

Even if your organization is not subject to CCPA, as of January 1, 2023, it may be subject to the amended version of the law, known as the California Privacy Rights Act (CPRA). Like CCPA, the amended law does not include specific IT security requirements, and the recovery for breach section remains the same with respect to the amounts. Notably, CPRA expands consumers' rights to include a right to correct data, a right to limit "sale" or sharing of "sensitive personal information", and a right to limit sharing of personal information. CPRA defines "sharing" to mean any disclosure (regardless of whether your organization receives anything in return) "to a third party for cross-context behavioral advertising." In addition, CPRA would also remove two common limited exceptions from CCPA: the "employee" exception and the "business-to-business" exception, although there is legislation pending as of this writing relating to those limited exceptions.

*Relevance to IT*: Even if your organization has outsourced the consumer request process to a third party, in order to comply with CCPA, IT still needs to know where the relevant consumer data is located—including whether it has been provided to third-party service providers. With respect to CPRA, the new "sharing" disclosure has led to a technological change—moving the advertising industry away from cookies. In addition, if the "employee" and "business-to-business" limited exceptions disappear as of January 1, 2023, then companies that had previously not had many obligations under CCPA may find that implementation should begin now. As described above, if your organization has personal data that the lawyers tell you is no longer needed, delete it—you don't have to provide a substantive response to consumer requests relating to information you do not have.

### 8.2.4 VIRGINIA, COLORADO, AND UTAH

Virginia enacted the Virginia Consumer Data Protection Act, which is scheduled to go into effect as of January 1, 2023. Subject to many exceptions, consumers will have rights similar to CCPA: the right to access personal data that the controller is processing, the right to correct data; the right to delete information provided by the consumer or about the consumer; the right to obtain a readily portable copy of the data; and the right to "opt out of the processing of the personal data for purposes of (i) targeted advertising, (ii) the sale of personal data, or (iii) profiling in furtherance of decisions that produce legal or similarly significant effects concerning the consumer." The new law also includes requirements relating to privacy policies and agreements with processors. It also requires that controllers must

> establish, implement, and maintain reasonable administrative, technical, and physical data security practices to protect the confidentiality, integrity, and accessibility of personal data. Such data security practices shall be appropriate to the volume and nature of the personal data at issue.

The Colorado Privacy Act is scheduled to be effective as of July 1, 2023, and the new law has some similarities with California's and Virginia's privacy laws. Consumers will have the right to opt out of a sale of their data, but also from targeted advertising and "profiling in furtherance of decisions that produce legal or similarly significant effects concerning a consumer."[8] Consumers will also have rights of access, correction, deletion, and data portability. Note that Colorado specifically requires data minimization: "A controller's collection of personal data must be adequate, relevant, and limited to what is reasonably necessary in relation to the specified purposes for which the data are processed."[9] (A "controller" determines the purposes for and means of processing "personal data".) The new law also contains some general security obligations:

> A controller shall take reasonable measures to secure personal data during both storage and use from unauthorized acquisition. The data security practices must be appropriate to the volume, scope, and nature of the personal data processed and the nature of the business.

Utah passed a privacy law in March of 2022, which is also similar to the previous states discussed. It goes into effect on December 31, 2023. Consumers will have the right to access, delete, and opt out of the sale their personal data or from targeted advertising. Similar to Colorado, Utah also requires data minimization. The new law also contains general security requirements: "Considering the controller's business size, scope, and type, a controller shall use data security practices that are appropriate for the volume and nature of the personal data at issue."[10]

Many other states are also considering legislation.

*Relevance to IT:* Ask your lawyer which requirements will apply to your organization. If California's privacy law does not already apply to your organization, work with your lawyer to determine how to come into compliance because there is no "one size fits all" solution.

## 8.2.5   PCI-DSS

If your organization accepts credit cards or debit cards as payment, you are probably already familiar with the Payment Card Industry Data Security Standard (PCI-DSS). PCI-DSS is set of standards created by the payment card industry (Visa, MasterCard, etc.), rather than a state legislature. Only Nevada requires PCI-DSS compliance by law.[11] These requirements are much more focused on IT security than the laws described above. The current version of PCI-DSS requirements—version 3.2.1—is 139 pages in length but summarizes the requirements in the table below[12]:

PCI Data Security Standard – High Level Overview

| Build and Maintain a Secure Network and Systems | 1. | Install and maintain a firewall configuration to protect cardholder data |
| | 2. | Do not use vendor-supplied defaults for system passwords and other security parameters |
| Protect Cardholder Data | 3. | Protect stored cardholder data |
| | 4. | Encrypt transmission of cardholder data across open, public networks |
| Maintain a Vulnerability Management Program | 5. | Protect all systems against malware and regularly update anti-virus software or programs |
| | 6. | Develop and maintain secure systems and applications |
| Implement Strong Access Control Measures | 7. | Restrict access to cardholder data by business need to know |
| | 8. | Identify and authenticate access to system components |
| | 9. | Restrict physical access to cardholder data |
| Regularly Monitor and Test Networks | 10. | Track and monitor all access to network resources and cardholder data |
| | 11. | Regularly test security systems and processes |
| Maintain an Information Security Policy | 12. | Maintain a policy that addresses information security for all personnel |

*Relevance to IT:* It is a common myth that PCI-DSS only applies to credit card processors. Instead, the standard states that it applies to "*all* entities involved in payment card processing—including merchants, processors, acquirers, issuers, and service providers. PCI-DSS also applies to *all* other entities that store, process or transmit cardholder data (CHD) and/or sensitive authentication data (SAD)."[13] As part of your mapping where personal data is within your organization and which vendors have it, be sure to include credit card data in your maps.

## 8.2.6   HIPAA

Until recently, the Health Insurance Portability and Accountability Act (HIPAA) and its privacy and security regulations were probably America's most comprehensive set of requirements. HIPAA is focused on personal health information, and its requirements focus on "covered entities," such as health care providers, and the organizations that assist them ("business associates"). HIPAA's security regulations contain requirements relating to administrative, physical, and technical measures, which are designed to be flexible enough for a national chain of hospitals and for a one-person doctor's office. This regulation also includes a separate set of security breach notification requirements, in addition to the state breach notification requirements discussed above. The privacy regulation also provides individuals with a right of access to some of their information, a right to amend the information, and a right to receive an accounting of the disclosures of their information.

*Relevance to IT:* If your organization is a "covered entity" under HIPAA, you are probably very familiar with its requirements because they have been in existence for 20 years. If your organization has a security incident, notice may also have to be provided to the U.S. Department of Health and Human Services. Health and Human Services may follow up with an audit, which will likely begin with a request to see your most recent risk assessment. Make sure you have it handy.

If your organization is not a "covered entity," but some customers are in the health care area, ask your lawyer if your organization has signed any Business Associate Agreements, that will place many of HIPAA's requirements on your organization.

### 8.2.7    FTC Act "Unfair or Deceptive Practices"

The law creating the Federal Trade Commission (FTC) was enacted in 1914, and the FTC has been making headlines with consents in the privacy and security area for the past 25 years. Although the FTC has specific regulations in a few areas, such as the Children's Online Privacy Protection Act, most of the consent agreements relate to the FTC's authority to regulate "unfair or deceptive practices or practices."[14] The FTC will frequently enter into a consent after a company has had a security breach, where the FTC will focus on the organization's privacy policy and, sometimes, advertisements and how the breach raises questions about the accuracy of those policies and advertisements. On March 15, 2022, the FTC enters into a consent agreement with a company that was accused of failing to investigate a hacking incident that the company had been notified by third parties affected personal data, of failing to use multi-factor authentication, and for over-retaining data. The FTC has proposed settling the matter for (1) $500,000, (2) the company's implementation of an information security program that is subject to ongoing third-party audits, and (3) recordkeeping obligations for 20 years.[15]

*Relevance to IT:* Especially if your organization offers apps or is otherwise offerings products or services that can collect personal information, review your organization's privacy policy. If you think it should change, talk to your lawyer—who will probably be updating the privacy policy anyway in light of all the new laws described above.

### 8.2.8    FERPA

The federal Family Educational Rights and Privacy Act (FERPA) relates to privacy of student educational records. If your organization is an educational institution (including colleges and universities) or provides goods or services to educational institutions that cause you to have access to student records, FERPA may apply. FERPA requires that educational institutions provide an annual notice to parents of students under 18 (or directly to students who are 18 or over) with an annual notice of their rights under FERPA. Those rights include the right to review the student's education records and to make corrections, to control the disclosure of certain information, and to comply to the Department of Education for violations. Although educational institutions may disclose certain "directory information" without consent, other disclosures require parental/student consent. "Directory information" includes such information as a student's name, degree, birth date, phone number, major, activities, and sports. The regulations contain additional exceptions.

*Relevance to IT*: If your organization collects personal data, are any of your organization's customers educational institutions? If so, speak with your lawyer to make sure your practices reflect what those educational institutions are including in their annual notices.

### 8.2.9   GDPR AND PERSONAL DATA BELONGING TO NON-US RESIDENTS

The European Union adopted the General Data Protection Regulation (GDPR) in 2016 (so the UK was included). These comprehensive regulations relate to residents of the European Union and include some principles for processing personal data:

- Lawfulness, fairness, and transparency
- Purpose limitation (you can use the data only for the stated purposes)
- Data minimization
- Accuracy
- Storage limitation
- Integrity and confidentiality
- Accountability

Requirements under GDPR include:

- Consent to disclosures and transfers
- Right to erasure ("right to be forgotten")
- Limits on automated decision-making
- Data breach notification.

If your organization has customers outside the United States, as you can see, very different rules can apply. In addition, many other countries have, or are considering, privacy laws, including Argentina, Australia, Brazil, Canada, China, Dubai, Hong Kong, India, Israel, Japan, Mexico, Russia, Singapore, and South Korea. Note that requirements can include "data localization"—the original data must stay within that country.

*Relevance to IT*: International compliance is the province of lawyers. Ask your lawyer what you need to do to comply.

### 8.2.10   CONTRACTUAL AGREEMENTS ON DATA HANDLING

Even though some of the laws described above have some requirements for agreements with third parties (including HIPAA's Business Associate Agreements and the

CCPA service provider requirements), many companies add their own requirements to their agreements. Those agreements may include annual security surveys, audit requirements, or breach notifications within, for example, 48 hours of discovery.

*Relevance to IT:* Although you may already be completing security surveys for customers, ask your lawyer if there are agreements with any other requirements that impact IT.

## 8.3   BIOMETRIC SECURITY

Biometric security systems are increasingly used by companies and government agencies for identity and access management, as well as for identifying individuals under surveillance. In addition to security, the driving force has been convenience: no need to remember passwords or carry security tokens or cards. With facial recognition, there is no direct contact with the individual.

Biometric security systems come in two varieties: physiological and behavioral. Physiological biometrics include facial recognition, fingerprints, finger geometry, iris recognition, vein recognition, and retina scanning. Behavioral biometrics systems include voice recognition, signature or keystroke dynamics, the way an individual uses objects, or the individual's gestures or gait (the sound of steps).

Facial recognition, for example, involves a camera that detects and locates the image of a face, either looking straight ahead or in profile. Next, an image of the face is captured and the geometry of the face is measured, key features being measurements of the distance between eyes, the depth of eye sockets, distance from forehead to chin, shape of cheekbones, and the contours of the lips, ears, and chin. These measurements create a set of digital information based on the person's facial features, creating a numerical code called a "faceprint." These faceprints, like fingerprints, are unique. In a typical biometric security system, faceprints of employees or other trusted individuals are stored and then automatically compared to the faceprint generated when a person seeks admission to a facility, and if the faceprints match, the system permits access to the individual. Some biometric security systems employ three-dimensional "liveness" technology, which helps prevent bots and bad actors from using stolen photos, injected deep fake videos, life-like masks, or other spoofs to gain admittance. These faceprints, really mathematical formulae, are known as "biometric data."

Protection of biometric data is of growing concern to privacy regulators around the world because it is so personal—unlike other personal data such as social security or account numbers, you can't change your biometrics. For that reason, biometric data is commonly included among the personal data elements in data breach laws, meaning that if it is accessed or acquired by a threat actor, it constitutes a data breach requiring notification of the individuals whose biometric data was affected. Biometric data is a listed personal data element in many state data breach laws, including California's Consumer Privacy Act, and New York's SHIELD Act, as well as the data protection laws in the European Union and the United Kingdom.

Several states have enacted specific laws regulating the use of biometric security systems, and many more states are moving toward enacting their own. The leading law is the Illinois Biometric Information Privacy Act (BIPA), passed in 2008, which:

requires informed consent prior to collection of an individual's biometric data, meaning that for use with employees or customers, those employees or customers must consent in advance to its use of their biometric data. BIPA also mandates protection obligations and retention guidelines; prohibits profiting from biometric data; creates a private right of action for individuals harmed by BIPA violations; and provides statutory damages up to $1,000 for each negligent violation, and up to $5,000 for each intentional or reckless violation. The statutory damages provision in BIPA have led to hundreds of lawsuits against businesses that failed to obtain individuals' consent for collection and use of their biometric data. Facebook, Google, and Amazon have also been named in BIPA lawsuits for their use of biometric data (e.g., Facebook's practice of tagging people in photos using facial recognition without their consent).

A particularly controversial biometric surveillance product is offered by Clearview AI, which creates biometric data from photographs scraped off the Internet, which is reported to include over three billion people in its biometric database. Clearview AI is popular with law enforcement and has been introduced at schools, to identify visitors who may present risk. Clearview AI has been the target of many lawsuits in Illinois (where it has stopped offering its software) and elsewhere. Most biometric security systems only store biometric data for employees or others who have a legitimate need to enter facilities. In such cases only the biometric information of those individuals are stored, and consent can readily be obtained in advance.

Texas and Washington have enacted biometric privacy laws which do not create private rights or action, but like Illinois they require reasonable care to guard against unauthorized access or acquisition of the biometric data, and retention of the data for no longer than necessary for the purpose for which it was collected. A best practice is to encrypt biometric data while stored, to delete it after it is no longer needed for its purpose, and to provide details of such security precautions to the affected individuals.

Another indicator of the likely proliferation of biometric privacy laws, New York City amended its administrative code in July 2021 to include a new regulation covering the use of biometric identifier information (BII) by "commercial establishments," which includes places of entertainment, retail stores, and food and drink establishments.[16] The new ordinance requires such commercial establishments to post a clear and conspicuous sign notifying customers of the biometric collection activity, and makes it unlawful to sell, lease, or otherwise profit from the BII. It also creates a private right of action for aggrieved individuals to sue for violations. Later in July, New York City passed the Tenant Data Privacy Act which places limits on the use of BII by "smart access" building owners.[17]

*Relevance to IT*: If implementing biometric security, understand that there are fewer legal risks under Illinois, Texas, and Washington state law if you restrict its operation to identification of employees and independent contractors only, and pay close attention to providing each of them with explicit notice, including in writing and within company policies, and obtain affirmative consent to the holding of their biometric information and photo. Particular attention should be paid to encryption of the biometric data files generated by the system and, to a lesser degree, the photos from which they were derived. If you need to use a biometric system that can identify random individuals, be aware that if the system sources biometrics for identifying information without first obtaining consent from the individuals involved, liability can follow in some states.

## 8.4   COLLECTING DIGITAL EVIDENCE AND ELECTRONIC DISCOVERY

### 8.4.1   Forensically Sound Collection of Digital Evidence

IT teams frequently are tasked to collect digital evidence on behalf of the organization's legal department. These requests come through the legal department for a number of reasons and can relate to internal investigations (e.g., sexual harassment allegations), regulatory or law enforcement investigations, or litigation where the organization is a party, or is subpoenaed for materials needed in a litigation in which it is not a party.

In some cases, this calls for IT personnel to search for the potential evidence; other times they are directed to collect a particular electronic file from the organization's data storage system or on an employee's personal device (e.g., workstation, laptop, or smartphone). The evidence sought could include e-mail, text messages, electronic documents, computer logs, or social media postings. Searching for such potential evidence across the network—electronic discovery is discussed below.

Legal will expect IT to collect digital evidence in a manner that ensures its admissibility in court. Before 2018, admission of digital evidence into evidence potentially required testimony in court by the IT personnel to describe the method by which they copied the file or e-mail at issue, as part of the evidence authentication process. In 2018, the Federal Rules of Evidence recognized the advancement of computer forensics and amended F.R.E. 902 ("Evidence That Is Self-Authenticating") by adding the following new section 902(14) ("Certified Data Copied from an Electronic Device, Storage Medium, or File"):

> Data copied from an electronic device, storage medium, or file, if authenticated by a process of digital identification, as shown by a certification of a qualified person that complies with the certification requirements of Rule 902(11) or (12). The proponent also must meet the notice requirements of Rule 902(11).

The Advisory Committee on Evidence, which authored the new section 902(14), explained the new rule as follows:

> Today, data copied from electronic devices, storage media, and electronic files are ordinarily authenticated by 'hash value'. A hash value is a number that is often represented as a sequence of characters and is produced by an algorithm based upon the digital contents of a drive, medium, or file. If the hash values for the original and copy are different, then the copy is not identical to the original. If the hash values for the original and copy are the same, it is highly improbable that the original and copy are not identical. Thus, identical hash values for the original and copy reliably attest to the fact that they are exact duplicates. This amendment allows self-authentication by a certification of a qualified person that she checked the hash value of the proffered item and that it was identical to the original. The rule is flexible enough to allow certifications through processes other than comparison of hash value, including by other reliable means of identification provided by future technology.

Many organizations' IT departments' toolkits include computer forensic tools that enable IT to make forensically sound copies of electronic files or e-mails from their networks, electronic communication systems, and portable devices, including personal smartphones and tablets. These tools create a hash of the target files, and then take exact copies of the files which it also hashes and confirms that the hash for the copy matches the hash for the original. They also generate reports on the forensics.

Thus, IT departments can meet the authentication requirement by using a recognized forensic tool utilizing hash values and accompanying it with a "certification" of a "qualified person." A qualified person is someone who knows how data systems operate, which can be a skilled member of the IT team, or a third-party consultant. The rules of evidence are not clear on what is required in a certification; it is prudent that a sworn affidavit be used. The affidavit should include:

- Background and qualifications of the affiant
- Detailed description of the evidence
- Description of the retrieval method and verification process
- State that affiant certifies that the evidence is genuine and unaltered
- Forensic reports and documentation about the collection
- Signature and notarization

*Relevance to IT:* Consider obtaining computer forensic software and arranging to have a staff member certified in its proper use, to ensure a forensic collection capability. Alternatively, engage a respected and certified third-party service provider that can handle collections when they arise. Note that if a file is simply copied using a non-forensic tool, it can reset the file's metadata, such as "date created" and "date last accessed."

## 8.4.2 ELECTRONIC DISCOVERY

Legal also relies on IT departments to find potential evidence needed for litigation or government or internal investigations. When responding to a "request for documents" seeking potentially relevant Electronically Stored Information (ESI) from another party in a litigation, or from a government authority, it may arrive in the form of a subpoena that describes the sorts of electronic documents or e-mails needed.. Lawyers convert the demands in the subpoena into operational search criteria, often in Boolean format. The search criteria typically call for e-mails or documents:

- Within a particular timeframe
- From identified "custodians" (i.e., employees or others with accounts on the system)
- Sent or received e-mails from identified individuals
- Containing particular search terms (e.g., "withdraw! w/5 account!")
- Particular formats (e.g., Word, Excel, PowerPoint, graphics, e-mail, text messages)

Search criteria, obviously, determine the limits of what files will be discovered, and their terms often are the subject of negotiation among the lawyers for the parties or government authorities involved. Your organization's lawyers may request that IT run alternative searches without collecting files, just to determine the number of "hits" each search obtains, information which can be useful for the purposes of negotiating the scope of ESI to be produced.

The organization's counsel may have a legal obligation to preserve ESI, even before it is collected from the network and connected devices. For example, if a company's counsel can "reasonably anticipate litigation," either a lawsuit to be brought by the organization, or one in which they can anticipate being a defendant, or can expect to be subpoenaed for ESI in a litigation or government investigation.[18] When legal determines the organization is obligated to impose a legal hold to preserve any ESI that relate to the litigation or investigation, it provides notice to all employees who are custodians of potentially responsive ESI that they must not delete any of it during the duration of the hold. Counsel should also provide written instructions to IT to make it impossible for identified custodians' ESI to be permanently deleted. This preservation of ESI may involve reaching back in time, to ESI created and stores months or years before the litigation hold goes into effect, and a strategy for preserving disaster recover media.

IT should ensure that it possesses adequate technical capabilities to ensure preservation of ESI under legal hold, to avoid the obligation to forensically collect all potentially responsive ESI on the system. It is common that the range of ESI preserved in a legal hold is much broader than that actually collected pursuant to instructions from counsel, and in the early stages of an investigation or litigation the search parameters may only be roughly understood by the lawyers.

While such searches can be conducted manually, organizations often employ electronic discovery ("e-discovery") tools capable of rapidly applying the search criteria to the files held by particular custodians and automatically performing a forensically sound collection of each. If the organization does not own its own e-discovery search-and-collection technology, outside consultants with expertise in such collections may be engaged. In either case, any e-mail or digital files must be authenticated pursuant to F.R.E. 902(14), as discussed above.

Organizations not only need to demonstrate that they conducted a thorough search for potentially relevant digital evidence, but also that they preserved all potential electronic evidence from deletion prior to the time of the search and collection. Such preservation, known as a "litigation hold" or a "legal hold," is required promptly after the organization becomes aware (or should have become aware) that it may be involved in litigation or an investigation. Hold notices typically are sent to each custodian (employees, independent contractors, or others with accounts on the network) informing them not to delete files or communications. IT, however, is counted upon to make complete deletion of such e-mails or files impossible. The hold should stay in place until your organization's lawyers instruct otherwise, including after the collection has occurred (because there may be the necessity for secondary collections).

The Federal Rules of Civil Procedure, which regulates discovery in federal court cases, establishes a procedure in Fed. R. Civ. P. 26(b)(2)(b) for instances in which the data sought are from "sources that the party identifies as not reasonably accessible because of undue burden or cost." The organization's lawyers will rely on IT to make them aware of particular search-and-collection challenges that may qualify as "not reasonably accessible."

It is important for IT to treat the collected e-discovery data with particular care—so as not to create legal issues or questions around the handling of the evidence in the matter—including drafting a written "chain of custody" memorandum (or form) describing exactly how the data was collected, the media onto which the data was loaded, and how it was delivered either to the organization's legal department, or to an outside party.

Data collected for an e-discovery project is typically loaded into an e-discovery review platform, which permits teams of lawyers to view the files regardless of their file type, as well as to conduct searches (for which the data is indexed) and tag individual files or messages as relevant or privileged. The data typically must be processed to create a load file for ingestion into the review platform. This may require IT to coordinate delivery of the collected e-discovery data to an outside vendor for such processing.

Once the collected ESI is processed and loaded into an attorney review platform, attorney reviewers tag responsive communications or files for production to the requester. Some responsive documents may be withheld from production if they are subject to the Attorney-Client privilege or the Attorney Work Product doctrine. Others may be withheld from production—or produced in redacted form—due to their confidential nature. This does not affect IT's e-discovery collection responsibilities, as the withheld documents must be identified to the requester.

*Relevance to IT:* The most serious legal consequences for an organization in an e-discovery project come from a failure not to preserve potentially relevant files. If there is some search-and-collection problem, it can be rectified unless the data has been deleted. IT should move promptly when advised by counsel to impose a legal hold and prevent employees from permanently deleting any potentially responsive e-mails or documents.

## 8.5  CRIMINAL LIABILITY FOR SURVEILLING EMPLOYEES

Circumstances sometimes make it necessary for organizations to surveil its employees' telephone conversations and electronic communications. This may arise in the context of government regulatory or criminal investigations, or an internal investigation, or everyday security or quality control reasons. When the surveillance is ordered or conducted by a government authority, its legality is governed under a body of law and judicial interpretations of the rights of surveilled individuals under the U.S. Constitution's Fourth Amendment.

There is a separate body of legal analysis governing non-governmental organizations, both under U.S. federal and state laws. The primary federal statute is the Electronic Communications Privacy Act of 1986 (the "ECPA"), which incorporates the Stored Wire Electronic Communications Act and an update of the Federal Wiretap Act of 1968 as Title 1 of the ECPA. *See* 18 U.S.C. Sections 2510–2520. The Wiretap Act creates criminal liability for any person who "intentionally intercepts, endeavors to intercept, or procures any other person to intercept or endeavor to intercept, any wire, oral, or electronic communication." 18 U.S.C. Section 2511(1). It also makes it illegal to use any electronic, mechanical, or other devices to intercept oral communications under specified circumstances, or to disclose or otherwise use the contents of illegally intercepted communications. 18 U.S.C. Section 2515.

There is, however, the courts recognize a limited exception from Wiretap Act liability for non-governmental organizations that intercept or monitor employees' communications in "the ordinary course of business." The exception calls on courts to balance the legitimacy of the employers' business justification for interception versus the employees' expectation of privacy for personal conversations.

The Wiretap Act dates to the 1980s, and for years most of the caselaw interpreting the limits of "the ordinary course of business" involved cases where employers were sued for monitoring employees' telephone communications (more recent cases apply the same principles e-mail and text messages). We will examine how the Act and judicial decisions have interpreted the scope of employers' "ordinary course of business" excepting interception of personal telephone calls from criminal liability under the Wiretap Act.

First of all, business calls clearly qualify for the exception. The challenge comes when employees' personal conversations are monitored (and thereby could have been "intercepted" under the statute). Courts tend to find the exception applies in situations where the employer has issued a written policy that either clearly identifies monitored phone lines, or expressly prohibits personal calls on monitored lines. Such policy provisions would reinforce the position that employees have a reduced expectation of privacy. To invoke this exception with respect to personal communications the employer must articulate a legitimate business interest in monitoring telephone conversations. For example, a federal appeals court held for the employer where it found that an employee's supervisor had particular suspicions that the employee was disclosing confidential information to a business competitor, had warned the employee not to disclose such information, and knew that a particular telephone call was with an agent of the competitor.[19] The appellate court held that the employer's listening in on the conversation (using interception equipment furnished by a communication services provider) was in the "ordinary course of business" and, thus, not a violation of the Wiretap Act.

Whether a court will agree with the employer that their monitoring of an employee's personal telephone call falls within the exemption to the Wiretap Act depends on the factual context. Other federal appellate courts have held against employers seeking to invoke the exception, based on the circumstances. One court held that the employer violated the Act by tape recording and listening to all calls, including personal calls, even though the employer's suspicions of theft would have been exempted from liability had the company merely monitored the calls to the extent necessary to determine their nature.[20] In another case, the employer had a policy of monitoring employees' sales calls, advising employees that their personal calls would not be monitored except to the extent necessary to determine that a call was of a personal nature. The employee brought the lawsuit after he discovered that the employer had monitored a call in which the employee discussed a job interview with a prospective employer. The court held that the interception was not in the ordinary course of business, stressing that the employee was an at-will employee and, therefore, the employer had no legal interest in his future employment plans, explaining that the exception cannot be extended to mean anything about which an employer is curious. In these circumstances, the court concluded that personal calls could not be intercepted in the ordinary course of business under the employer's policy, except to the extent necessary to determine whether the call was personal or not.[21]

Under the Wiretap Act, and under the laws of most U.S. states,[22] there is no liability under if the organization has obtained the consent of one party to a telephone conversation who has full knowledge that the communication will be recorded. Other states[23] clearly (or potentially) require consent from all parties to a telephone conversation under some or all circumstances.

In order to qualify for the "ordinary course of business" exclusion under the Wiretap Act, an employer also is required to perform the monitoring only using standard telephone-related equipment supplied by the service provider for connection to the phone system.[24] Courts have found the exclusion not to apply when it finds that the monitoring was performed using telephone equipment that is not normal and was not installed by a standard service provider.[25]

Other legal considerations to keep in mind. In many states, consent requirements only apply in situations where the parties to the telephone conversation have a reasonable expectation of privacy. What constitutes "consent" can vary under the laws of different states in terms of whether that consent must be explicit or can be implied from the circumstances. Because it can be difficult to know exactly which states' laws apply to the monitoring or recording of a particular telephone conversation (consider that you may not know a party's residency or location at the time of the call), it is a best practice to set a procedure that observes the most restrictive state laws, and make exceptions to that procedure only after consulting with legal counsel.

Many of the legal concepts governing monitoring of employee telephone calls also apply to other electronic communications including e-mail. But there are differences. To begin with, the ECPA defines "electronic communication" as any electronic messages currently in transmission. Once an e-mail, text, or other electronic communication has been sent, the transmissions become "electronic storage," which courts have determined employers can monitor when stored on the organization's information systems. The definition of "intercept" in the Wiretap Act is unsettled with respect to e-mail and electronic communications, but likely applies to obtaining a communication in real time (like a wiretap) as opposed to accessing a stored communication like an e-mail or text message in storage. Access to stored communications falls under the Stored Communications Act,[26] which is within the Wiretap Act. The Stored Communications Act does not prevent an employer from accessing communications if they are stored on an employer-provided wire or electronic communications services if done in a manner consistent with the employer's own policies, on condition that such policies are clearly disclosed to employees. However, if the employer accesses communications stored elsewhere, for example, an employee's personal social media account or personal e-mail account with a third-party service provider, it could face civil or criminal penalties unless it first obtains the employees' permission. Be aware that many states prohibit employers from requiring or requesting that an employee verify a personal online account (e.g., social media accounts) or provide log-on information to personal accounts.

Courts have also held that employees have little expectation of privacy while on company grounds or working with company equipment, including company computers and vehicles. However, organizations must be careful with respect to employee's personal communications with e-mail and other electronic communications, as with telephone communications. The company must have a legitimate reason for monitoring and may allow some additional scope monitoring if the employee has given consent, but consent does not usually apply to monitoring of personal communications. To obtain employee consent, the organization can require employees to acknowledge and accept their monitoring practices at the time of onboarding, even before logging into company devices or network.

Four U.S. states—California, Florida, Louisiana, and South Carolina—explicitly state a right to privacy in their constitutions, and some states have laws that potentially pose a legal risk when monitoring personal e-mail or other electronic communications without receiving consent to all parties to the communication. For example, California Penal Code Section 632(a) makes it a crime if someone

> uses an electronic amplifying or recording device to eavesdrop upon or record the confidential communication, whether the communication is carried on among the parties in the presence of one another or by means of a telegraph, telephone, or other device....

Thus, if a company reviews a personal e-mail from an employee's account without consent from the sender and all recipients, it may be vulnerable to a lawsuit if one of them is a California resident. Some companies have attempted to remedy this issue by posting notice on their websites that there should be no expectation of privacy by anyone communicating with its employees on company systems.

New York recently enacted an amendment to its Civil Rights Law that requires employers to provide notice to employees before monitoring or intercepting and employee's telephone conversations, e-mails, Internet access, or Internet usage.[27] Employers must give written or electronic notice to all employees susceptible to monitoring upon hiring, and also post the notice in a conspicuous place. The law prescribes a $500 fine for a first offense, $1,000 for the second offense, and $3,000 for a third and all subsequent offenses. It does not create a private right of action enabling an employee to sue the employer for a violation.

*Relevance to IT:* Monitoring of employee telephone calls, e-mails, or other electronic communications brings legal risks unless the employer tailors their methodology to meet U.S. federal and state laws. Note that courts balance the legitimacy of the employer's business justification for interception against the employee's expectation of privacy for personal communications. This includes providing clear notice to employees of potential monitoring of communications, including it in policies that are consented to by the employee upon hiring, with occasional reminders. Monitoring protocols must include procedures to halt monitoring when it becomes apparent that the communication is personal in nature and be cognizant that some states require that other non-employee parties to the communication also consent to monitoring. And when monitoring telephone conversations, be aware that the Wiretap Act requires that only standard telephone company equipment be used.

## 8.6  ORGANIZATIONAL SECURITY, FINANCIAL INSTITUTIONS, AND CRITICAL INFRASTRUCTURE

### 8.6.1  GRAMM-LEACH-BLILEY ACT

The federal Gramm-Leach-Bliley Act of 1999 was enacted in part to enable sharing of personal information between financial institutions, while protecting privacy. The law requires an annual notification and California has some additional requirements under its state law relating to sharing information with affiliates. The law requires that financial institutions provide privacy notices to consumers. The regulations require that financial institutions must "develop, implement, and maintain a comprehensive information security program."[28]

*Relevance to IT:* The regulations do not provide specific requirements, so you should consult with your lawyer to decide what level of security would be appropriate for the size and nature of your organization as well as the types of information collected.

### 8.6.2  SARBANES-OXLEY

Enacted to address the accounting scandals of Enron and others, the law named the Public Company Accounting Reform and Investor Protection Act of 2002 is more commonly referred to by the names of its sponsors, Senator Paul Sarbanes and Representative Michael Oxley, abbreviated as "SOX." Among the many provisions of SOX, two provisions focus on publicly traded companies maintaining the internal controls necessary for reasonable assurance regarding reliability of companies' financial reporting and preparation of financial statements in accordance with generally accepted accounting principles: Section 302[29] and Section 404.[30] Section 302 requires publicly traded companies to have internal procedures designed to provide accurate financial statements, and requires Chief Executive Officers and Chief Financial Officers to certify that they have designed such internal controls. Section 404 requires publicly traded companies to provide a report containing an assessment of the "effectiveness of the internal control structure and procedures of the issuer for financial reporting" in accordance with SEC rules.

Publicly traded companies cannot simply delegate these obligations by outsourcing significant processes to third parties. In addition, if a publicly traded company outsources a significant process that relates to controls affecting its financial statements, management is required to obtain assurances about the controls in place for that service organization.[31] The SEC's Guidance states that, if management is unable to determine the effectiveness of the service organization's controls, "management must determine whether the inability to assess controls over a particular process is significant enough to conclude in its report that ICFR [internal control over financial reporting] is not effective."[32] Consequently, even companies that are not publicly traded can be affected by SOX if they work with publicly traded companies.

*Relevance to IT:* One common way publicly traded companies demonstrate controls is use of a Statement on Standards for Attestation Engagements (SSAE-18) for the company, and to require an SSAE-18 for each service provider that may also affect financial reporting.

### 8.6.3 STATE REGULATORY REQUIREMENTS, INCLUDING NY DFS CYBERSECURITY REGULATION

The New York State Department of Financial Services (DFS) promulgated a detailed cybersecurity regulation, which went into effect in March of 2017.[33] The regulation affects financial institutions licensed by DFS, including banks, securities brokerages, and insurers. The requirements include maintenance of "a cybersecurity program designed to protect the confidentiality, integrity and availability of the Covered Entity's Information Systems." The covered entity must conduct a periodic risk assessment and have in place several security policies. The regulation requires annual penetration tests, and bi-annual vulnerability assessments. It also requires audit trails, and implementation of multi-factor authentication, as well as policies relating to service providers. Significantly, DFS requires an annual certification of compliance and prohibits certification unless the organization is in compliance. If a covered entity has a security incident, it must notify DFS within 72 hours from the time the entity determines that a cybersecurity event has occurred.

On November 18, 2021, the U.S. federal banking regulators Office of the Comptroller of the Currency, Federal Reserve Board, and Federal Deposit Insurance Corporation jointly announced a final rule[34] that will, as of May 1, 2022, require banking organizations (which includes the U.S. operations of foreign banking organizations) to notify their regulators as soon as possible but no later than 36 hours of identifying a significant "computer-security incident" that results in "actual harm" and rises to the level of a "notification incident" as defined in the final rule. The proposed rule would also impose a separate notification requirement on companies (such as data processing companies) that provide certain services to those banks. Those service providers would be required to notify

> each affected banking organization customer as soon as possible when the bank service provider determines that it has experienced a computer-security incident that has caused, or is reasonably likely to cause, a material service disruption or degradation for four or more hours.

With respect to insurance, the National Association of Insurance Commissioners adopted a model law, which 19 states have adopted as of this writing, although with some variations. The National Association of Insurance Commissioners model is similar to the DFS cybersecurity regulation, including the 72-hour requirement for security incident notification. As an example of a variation, Indiana requires notice within three business days.

Finally, on February 9, 2022, the SEC proposed "Cybersecurity Risk Management Rules and Amendments for Registered Investment Advisers and Funds." The SEC announced amendments on March 9. The proposed rule would require advisers and funds to disclose certain cybersecurity risks and incidents to current and prospective clients. The SEC proposed rule would require advisers to promptly, but no later than 48 hours, report significant cybersecurity incidents to the SEC, including on behalf of itself and on behalf of a client that is a registered investment company or business company or a private fund. The 48-hour clock starts as soon as the adviser has a "reasonable basis to conclude" that a significant incident has occurred or is occurring.

*Relevance to IT:* If your organization is a financial services organization, or has customers in those industries, the very quick notification requirements may be of special concern. Your lawyer may be able to help you identify which organizations are in those industries.

### 8.6.4 FEDERAL CYBERSECURITY FRAMEWORK FOR CRITICAL INFRASTRUCTURE

The National Institute of Standards and Technology (NIST), which is part of the U.S. Department of Commerce, has promulgated a voluntary Framework for Improving Critical Infrastructure Cybersecurity.[35] The risk-based framework can be applied to both government and private industry to improve critical infrastructure cybersecurity. The framework consists of five different functions: identify, protect, detect, respond, and recover, which functions include many detailed subsections. In February of 2022, the U.S. Government Accounting Office reported to Congress that only three of the 16 critical infrastructure sectors have adopted the framework.[36] In addition, beginning in 2021, certain owners and operators of critical infrastructure have received directives from the Department of Homeland Security relating to cybersecurity, including requirements for rapid reporting of security incidents.

*Relevance to IT:* Large organizations in private industry seem more amenable to the NIST framework, and some regulators recognize compliance with the framework as an acceptable standard. If you have not reviewed the framework, do so now to determine whether it would be a good fit for your organization. If your organization received a directive from Homeland Security (or some agency within DHS, such as the Transportation Security Administration), you should already have processes and procedures in place.

### 8.6.5 DEFENSE FEDERAL ACQUISITION REGULATION SUPPLEMENT (DFARS)

If your organization is a government defense contractor (or subcontractor), special additional requirements apply in addition to the NIST SP-800 series. If the organization has a current government defense contract, the organization must submit a current Cybersecurity Maturity Model Certification showing its cybersecurity maturity level—and must require such a Cybersecurity Maturity Model Certification from its subcontractors.[37] In addition, federal regulations also require that the organization had "adequate security on all covered contractor information system."[38] That regulation also requires that the organization "rapidly" report any incident directly to the Department of Defense. In addition, on October 6, 2021, the U.S. Department of Justice announced that it would use the False Claims Act to pursue government contractors that "put U.S. information or systems at risk by knowingly providing deficient cybersecurity products or services, knowingly misrepresenting their cybersecurity practices or protocols, or knowingly violating obligations to monitor and report cybersecurity incidents and breaches."[39]

*Relevance to IT:* NIST frequently updates and offers new drafts. The importance of keeping up-to-date with NIST and other government contractor standards has increased with the announcement of potential False Claims Act penalties if security is a material part of the goods or services sold to the federal government.

## NOTES

1. This chapter was contributed by Patrick J. Burke, Counsel, Norton Rose Fulbright US LLP. And Susan Ross, Senior Counsel, Norton Rose Fulbright US LLP.
2. Compare, for example, California's breach law (Cal. Civ. Section 1798.80 et seq.) with New York's (N.Y. Gen. Bus. Law Section 899aa).
3. 201 CMR Section 17.03(1).
4. N.Y. Gen. Bus. Law Section 899bb.
5. EyeMed Vision Care, Inc., Assurance 21-071 (Jan. 18, 2022), available at https://ag. ny.gov/sites/default/files/eyemed_aod_-_final_-_fully_signed.pdf.
6. Can you name the "two guys in a garage"? The answer is Bill and David, but you may recognize the company name from their surnames: Hewlett and Packard.
7. Cal. Civ. Section 1798.150(a)(1).
8. Colo. Rev. Stat. Section 6-1-1306(1)(A)(1).
9. Colo. Rev. Stat. Section 6-1-1308(3).
10. Utah Rev. Stat. Section 13-61-302(2)(b).
11. Nev. Rev. Stat. Section 603A.215.
12. Payment Card Industry Data Security Standard, "Requirements and Security Assessment Procedures," version 3.2.1 at 5 (2018).
13. *Id*. (emphasis in original).
14. 15 U.S.C. Section 45(a).
15. In the Matter of CafePress, available at https://www.ftc.gov/legal-library/browse/cases-proceedings/1923209-cafepress-matter?utm_source=govdelivery
16. NYC Admin. Code Sections 22-1201–1205.
17. NYC Admin. Code Sections 26-3001–3007.
18. See Fed.R.Civ.P. 37(e):
    If electronically stored information that should have been preserved in the anticipation of litigation is lost because a party failed to take reasonable steps to preserve it, and it cannot be restored or replaced through additional discovery, the court:
    (1) upon finding prejudice to another party from loss of the information, may order measures no greater than necessary to cure the prejudice;
    (2) only upon finding that the party acted with the intent to deprive another party of the information's use in the litigation may:
        (A) presume that the lost information was unfavorable to the party;
        (B) instruct the jury that It may or must presume the information was unfavorable to the party; or
        (C) dismiss the action or enter a default judgment.
19. See Briggs v. American Air Filter Co., Inc., 630 F.2d. 414 (5th Cir. 1980).
20. See Deal v. Spears, 980 F.2d 1153 (8th Cir. 1992).
21. See Watkins v. L.M. Berry & Co., 704 F.2d 577 (11th Cir. 1983).check underline with MS
22. Alabama, Alaska, Arizona, Arkansas, Colorado, District of Columbia, Georgia, Hawaii, Idaho, Indiana, Iowa, Kansas, Kentucky, Louisiana, Maine, Minnesota, Mississippi, Missouri, Nebraska, New Jersey, New Mexico, New York, North Carolina, North Dakota, Ohio, Oklahoma, Rhode Island, South Carolina, South Dakota, Tennessee, Texas, Utah, Virginia, West Virginia, Wisconsin and Wyoming.
23. California, Connecticut, Delaware, Florida, Illinois, Maryland, Massachusetts, Michigan, Montana, Nevada, New Hampshire, Oregon, Pennsylvania, Vermont, and Washington.

24  See 18 U.S.C. Section 2510(5)(a)(i).

25  See, e.g., Deal v. Spears, 980 F.2d 1153 (8th Cir. 1992) (holding that exclusion did not apply where employer purchased a recorder at Radio Shack and privately connected the recorder to an extension phone line to automatically record all conversations); Sanders v. Robert Bosch Corp., 38 F.3d 736 (4th Cir. 1994) (a reel-to-reel tape recorder that continuously recorded certain telephone lines did not qualify for the exclusion, since it did not further the plant's communication system).

26  Stored Wire and Electronic Communications and Transactional Records Access, 18 U.S.C. Sections 2701–2713.

27  A.430/S.2628, N.Y. Civ. Rights Law Section 52-c.2(a). "Employer" includes "any individual, corporation, partnership, firm, or association with a place of business in the state" but does not include "the state or any political subdivision of the state."

28  16 C.F.R. Section 314.3(a).

29  15 U.S.C. Section 7241.

30  15 U.S.C. Section 7262.

31  Office of Chief Accountant, Div. of Corp. Finance, SEC, "Management's Report on Internal Control Over Financial Reporting and Certification in Exchange Act Periodic Reports Frequently Asked Questions," located at www.sec.gov/info/accountants/controlfaq/004.htm (last visited Jan. 2. 2008) (question 14).

32  Commission Guidance Regarding Management's Report on Internal Control Over Financial Reporting Under Section 13(a) or 15(d) of the Securities Exchange Act of 1934," Release Nos. 33-8810; 34-55929, Fed. Reg., Vol. 72, No. 123 at 35334 (June 27, 2007), Fed. Reg. at 35334.

33  23 N.Y.C.R.R. Part 500.

34  https://www.federalreserve.gov/newsevents/pressreleases/files/bcreg20211118a1.pdf

35  Available here: https://www.nist.gov/cyberframework.

36  GAO, "Critical Infrastructure Protection: Agencies Need to Assess Adoption of Cybersecurity Guidance," GAO-22-105103 (Feb. 2022), available at https://www.gao.gov/products/gao-22-105103?utm_campaign=usgao_email&utm_content=daybook&utm_medium=email&utm_source=govdelivery

37  DFARS Section 252.204-7021.

38  DFARS, Section 252.204-7012.

39  U.S. Dept. of Justice, "Deputy Attorney General Lisa O. Monaco Announces New Civil Cyber-Fraud Initiative," Oct. 6, 2021, available at https://www.justice.gov/opa/pr/deputy-attorney-general-lisa-o-monaco-announces-new-civil-cyber-fraud-initiative.

# 9 Cyber Security and Digital Forensics Careers

*In the middle of difficulty lies opportunity.*

—*Albert Einstein*

## 9.1 INTRODUCTION

Due to lack of training and qualifications of information technology and computer forensic investigation personnel, Julie Amero, a substitute teacher in Connecticut, was wrongly convicted of four counts of risk of injury to a minor. She lost her career and had her life turned upside down due to a malicious spyware application and the incompetence of security "professionals." The spyware was running on the classroom computer causing pornographic images to be shown to the students.

How did this happen? Julie innocently checked her personal e-mail using that classroom computer, left the room briefly, and upon her return saw, as did a few students, the pornography on the computer screen. The pornography pop-ups[1] were caused by spyware inadvertently installed when another user of that classroom computer downloaded a Halloween screen saver. Because of the school's amateur IT administrator, overreaction from a school principal, faulty forensic examination of the physical evidence, and false testimony from a computer forensics "expert," she was prosecuted and convicted (later over-turned) of risk of injury to a minor.[2]

What we can take away from this case is the importance of having a *qualified* computer forensics[3] examiner acquiring and analyzing evidence in addition to having a *qualified* information security professional protecting the critical assets of the enterprise. This includes training the employees on the proper use of the company computers as well as what to do when an *incident occurs*. All of these topics are discussed in this book. In this chapter, the focus is on the numerous career opportunities in the field of information and cyber security as well as a description on how to become a qualified professional in this exploding field.

## 9.2 CAREER OPPORTUNITIES

A job search can be overwhelming—especially in a technical field as there are mounds of available opportunities. In addition, with cyber security and digital forensics there are different certifications and job titles for similar positions. Here are a few pointers that will save you time when job searching in this field:

1. *Make your search general enough to include a variety of opportunities:* There are many different job titles for the same job.

DOI: 10.1201/9781003245223-9

2. *Familiarize yourself with multiple certifications:* A certification is obviously a plus; accordingly, some of the positions that require certifications may allow you to earn the certification within the first year of employment rather than having it at the start. Therefore, you could start looking for a job at the same time that you are working on the certification. If you already have a certification, note that the certifications advertised in a job description may be similar to the one you have. This will become clearer after you review the certification options later in this chapter. In addition, if you earned a degree that covers the tasks or knowledge domains listed in the job posting, the employer may not require a certification.

3. *Determine if the position requires security clearances, fingerprinting, and/or polygraph tests:* This requirement will be noted in the job description and the employer will most likely provide the means to accomplish that requirement.

As I am sure you are aware, and probably one of the reasons you picked up this book, there are many available opportunities in this field. The first thing to determine is your career interests and goals and then the qualifications required to get your foot in the door. The information in this chapter will facilitate that process by providing an overview of the tasks, training, and the necessary knowledge to acquire these positions.

This chapter is by no means an exhaustive review but is an excellent starting point to make sense of the immense amount of information out there regarding the cyber security and digital forensics professions. The first challenge you will encounter is sorting through the many job titles of these positions. Job titles vary as in many fields. Consider for example, a sheriff in one town means something different in another. A way to avoid the confusion is to focus more on the job description rather than the title.

Here are some of the MANY job titles you will come across during your search in the security field:

- *Information security job titles:* information security risk specialist, information security officer, information security specialist, information security analyst, data security specialist, information security architect, information security engineer, firewall engineer, malware analyst, network security engineer, director of security, security operations analyst, vulnerability researcher/exploit developer, security auditor, disaster recovery (DR)/business continuity (BC) analysis manager, data warehouse security architect, and penetration testing consultant
- *Digital forensic job titles:* emergency response managing consultant, computer forensics analyst, digital forensics technical lead, digital forensics engineer, cell phone forensics analyst, IT systems forensic manager, information security crime investigator/forensics expert, incident responder, computer crime investigator, intrusion analyst, and system, network, web, and application penetration tester

The purpose of each career outline coming up in this chapter is to give you an idea of what that professional may be asked to do or know. There is definite overlap in some of the tasks for the jobs listed. For example, you will note that the information security field includes an understanding of computer forensics knowledge. This is because the information security professional has designed and implemented the infrastructure that the computer forensics professional is investigating when an incident occurs. The information security professional needs to understand that it is not only important to implement a secure environment but also to implement effective monitoring, logging, and surveillance so that when (not if) the inevitable incident occurs, the computer forensics professional(s) will be able to analyze the system data to determine what happened to facilitate the prevention of the next occurrence. Thus, the computer forensics professional will have the necessary skill set to determine what has been compromised and, more important, be able to identify, recover, analyze, and preserve evidence in a forensically sound manner so that it will be admissible in court if the incident turns out to be a criminal offense. This may not be determined until all the data are analyzed.

### 9.2.1   A Summarized List of "Information Security" Job Tasks

1. *Develop and maintain the company security policy:* Create an acceptable use policy (AUP) to reduce the potential for legal action from the users of the system. The AUP is a set of rules applied that restrict the way the network may be used and monitored. For example, part of the AUP will address *general use and ownership* and will contain a statement similar to the following:

    While XYZ's network administration desires to provide a reasonable level of privacy, users should be aware that the data they create on the corporate systems remain the property of XYZ. Because of the need to protect XYZ's network, management cannot guarantee the confidentiality of information stored on any network device belonging to XYZ (SANS Institute).

2. *Monitor compliance with information security goals, regulations, policy, and procedures:* This requires knowledge of industry standards: Health Insurance Portability and Accountability Act (HIPAA), Payment Card Industry Data Security Standard (PCI-DSS), Federal Information Security Management Act (FISMA), and North American Electric Reliability Corporation-Critical Infrastructure Protection (NERC-CIP). For example, if you are working for an organization that deals with electronic health information (e.g., health plans, healthcare providers), then this National Institute of Standards and Technology (NIST) publication on HIPAA should be followed: "An Introductory Resource Guide for Implementing the Health Insurance Portability and Accountability Act (HIPAA) Security Rule." The HIPAA Security Rule focuses on safeguarding electronically protected health records. Thus, all healthcare and partnering organizations and anyone creating, storing, and transmitting protected health information electronically need to comply. The 117-page document focuses on improving the understanding of HIPAA overall, understanding the security concepts, and refers readers to other relevant NIST publications to assist in the compliance effort (Scholl et al. 2008).

Other regulations necessary to understand are *Sarbanes-Oxley (SOX) Act of 2002* (for public companies to secure the public against corporate fraud and misrepresentation) and the Gramm-Leach-Bliley (GLB) Act, which protect the privacy of consumer information held by financial institutions. Also see "monitor compliance" in the digital forensics task list later in this chapter.

3. *Security solutions development:* Design, deploy, and support the logical and physical security infrastructure for the network to safeguard intellectual property and confidential data. The starting point for the design and development can be accomplished by developing a security reference architecture that is essentially a template or blueprint to guide the security needs of the organization, including the major actors and activities. For example, the reference architecture could provide a consistent vocabulary of terms, acronyms, and definitions. This provides a common frame of reference during communication, thus facilitating the understanding of requirements among stakeholders. Figure 9.1 shows an example of a conceptual reference model for cloud computing.

4. *Investigate the security design and features of relevant information and security products necessary to deploy the security solution:* This includes supporting technologies such as intrusion detection systems (IDS), intrusion prevention systems (IPS), security logging, public-key infrastructure, data loss prevention, firewalls, remote access, proxies, and vulnerability management.

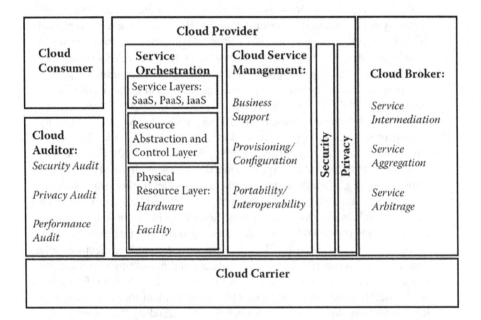

**FIGURE 9.1**  A conceptual reference model. (Modified from Liu, F. et al., 2007, NIST publication 500-292, http://www.nist.gov/customcf/get_pdf.cfm?pub_id=909505.)

5. *Maintain information and security products:* This task would include optimization, software upgrades, software patch installations, hardware upgrades, and diagnosis and resolution of software and hardware issues.

6. *Monitor and optimize system logs:* Review usage levels and performance; report misuse and security breaches. Provide weekly, monthly, and quarterly reports.

7. *Perform risk assessment:* This task addresses potential vulnerabilities and anticipates threats. The vulnerability assessment may be accomplished via a penetration test (aka pen-test) and/or a security audit. A pen-test is a way of testing the security of your system by simulating an attack. A security audit includes looking at all assets such as laptops, printers, and routers and performing, for example, vulnerability scans to assess the system for patch levels, open ports, etc. This essentially includes any activities that help determine whether the current configuration effectively mitigates security risks. Also see "perform risk assessment" of the computer forensics task list later.

8. *Maintain security incident handling process/plans:* This overlaps with the computer forensics profession. See "manage crisis/incident response" of the computer forensics task list later.

9. *Conduct or be involved in the incident response (IR) activities:* Be a member of the IR team and assist in an incident investigation.

10. *Facilitate security awareness and training:* Training is important to educate and drive the implementation and standardization of the company's security program.

11. *Create and maintain the business continuity (BC) and disaster recovery (DR) plans:* In the event of a disruption to your business due to anything from a malfunctioning system upgrade to an earthquake, your DR plan will describe the process by which your company can resume business activities after this planned or unplanned downtime. The BC plan will delineate how to keep your company functioning during this downtime.

## 9.2.2 A Summarized List of "Digital Forensic" Job Tasks

1. *Participate in e-discovery cases and digital forensics investigations:* Discovery is where each party involved in a lawsuit requests information from the opposing party. This information gathered will potentially be used in a trial. When the request is for electronic information such as Word documents, spreadsheets, e-mail, audio, and video, it is referred to as e-discovery. This requires following a forensically sound process to acquire, preserve, and analyze vast amounts of data from a variety of media types in addition to monitoring the chain of custody to protect the data against alteration and damage. Reports and presentations will also need to be created for possible inclusion in legal or policy disputes. Digital forensics investigations require analysis of the data in its entirety—not just logical files. This would include log files, swap space, slack space, and deleted files. Further details of the digital forensic process and the differences between e-discovery and digital forensics will be covered later in the book.

2. *Perform data recovery services for users:* Not all tasks are related to criminal activity. There are times when data recovery is needed due to human error such as file corruption or accidental deletion. A digital forensics professional will be able to recover data on any media including existing files, deleted yet remaining files, hidden files, password-protected files, encrypted files, fragmented data, and corrupted data to ensure that company information is retained.

3. *Create, evaluate, and improve the effectiveness of incident response (IR) policies:* You may be working with the information security professional to maintain the security incident handling processes and plans. NIST defines the IR plan as providing the "organization with a roadmap for implementing its incident response capability" (Cichonski et al. 2012). This book covers IR. NIST also recommends that the IR plan include elements that incorporate management support such as:

    a. A mission statement

    b. Strategies and goals to help determine the structure of the IR capability

    c. Senior management approval of the IR plan

    d. Determining the organizational approach to IR

    e. Determining how the IR team will communicate with the rest of the organization

    f. Metrics for measuring the IR capability to make sure it is fulfilling the goals

    g. Performing an annual review of the IR road map for maturing the IR capability

    h. Determining how the IR program fits into the overall organization

4. *Manage crisis/incident response:* Respond and confirm that there is in fact a problem (knowing the signs of an incident), analyze, contain, eradicate, and recover from the incident as well as perform an after-action report to determine lessons learned.

5. *Be familiar with the legal aspects of digital forensics as well as case law:* With the ease of Internet access and acquiring computing devices, there has been an obvious increase in crimes involving digital evidence. As a result, this has increased the need for digital forensics professionals. These professionals need to be very familiar with the law (the legality surrounding acquiring digital evidence in a criminal investigation) and case law (laws based on judicial decisions given in earlier cases). For example, in the case *US v. Carey*, there was a search warrant[4] to search the defendant's computer for drug-related evidence. The agent discovered child pornography and continued to search the computer for child pornography evidence. That additional searching exceeded the scope of the search warrant. Therefore, that search was considered an unconstitutional "general search." This restriction, based on the Fourth Amendment, which guards against unreasonable searches and seizures, caused the child pornography evidence to be suppressed from the case.

6. *Monitor the network:* This task overlaps with information security tasks. It would include the use of IDS, IPS, and/or network intrusion detection.

In addition, creating IDS/IPS signatures (patterns of common attacks or threats) where the IDS/IPS will detect network behavior patterns that match a known attack signature may also be included in the security posture of the system. This task may also involve deep packet inspection. Packet filtering is where the data or content of a packet is inspected for threats, and it includes applying content-filtering rules, creating statistics, and routing traffic. It is important for a digital forensics professional to be included in this process or at least in configuring these systems as the data collected will help in an investigation.

7. *Perform risk assessment:* Risk assessment also overlaps with the information security professional tasks. The risk assessment will determine which threats can exploit vulnerabilities that would damage company assets. Essentially, the results of the risk assessment should help to determine or improve the company's overall security posture. In other words, the risk assessment should result in recommendations of best practices for prevention of future threats. This task may also include pen-testing, which would intend to exploit known vulnerabilities, thus confirming risk. This would also require analysis of malware, attacker tools, and even reverse engineering of threats.

8. *Monitor compliance:* Compliance does not guarantee a secure infrastructure but it may lessen the odds of a breach. In addition to what was discussed in the information security task list, the computer forensics professional also needs to understand *breach laws.* Any company storing and accessing private consumer data is required to notify its consumers when their personal information may have been breached. Due to some states varying the law, senators are attempting to pass a national law to create a single national standard (Schwartz 2012). If an agency does not "take reasonable measures to protect and secure data in electronic form containing personal information," it could face a hefty fine per incident. Table 9.1 shows six large data breaches in 2021 (Bluefin 2021).

There are two more important skills/tasks required to be successful in this field, the first of which is communication. Security is a complex topic; however, at times it needs to be explained to a nontechnical person. For example, the results of a forensic analysis may need to be communicated to a lawyer, judge, or law enforcement official who may not have a technical background and who may need to merge your analysis results with the nontechnical aspects of a case. The point is to know your audience. For example, if you are presenting a forensics analysis to the CEO, you do not need to explain the log file settings. An overview of what the log files showed would suffice.

The second skill/task is project management. Every upgrade, implementation, and investigation is a project, which requires managing people, time, and probably a budget. You do not need a PMP (project management professional) credential, but it would be beneficial to take a course or read a book in the area of project management.

There are an abundance of opportunities for training in the fields of cyber security and digital forensics to obtain knowledge to perform the tasks discussed here. For

**TABLE 9.1**

**Six Big Data Breaches of 2021**

| Company | Amount of Data/# of Records | Breach |
| --- | --- | --- |
| 1. Colonial Pipeline | 100 gigabytes of data | Data stolen/ransom (2.3 million in Bitcoin) caused a disruption in the supply chain and an increase in gas process. |
| 2. Facebook, Instagram, and LinkedIn | 214 million | Unsecured database caused social media user personally identifiable information (PII) data leak. |
| 3. Bonobos (clothing retail) | 7 million | PII data leak. |
| 4. Kroger | 1.47 million | Third-party cloud provider (Accellion) leaked customer pharmaceutical records. |
| 5. Parler | 70 terabytes of data | Flawed scrapping API exposed 99% of posts, messages, video, user metadata. |
| 6. Volkswagen and Audi | 3.4 million | Customer and prospect data in Canada and the US. |

example, some colleges and universities are offering associate, bachelor, and graduate degrees in these professions. If you already have a degree in another area, a faster (and less expensive) way to break into this field is to earn a certification. Many training courses offered by professional or training organizations to help prepare for the certifications are outlined in the next section.

## 9.3   CERTIFICATIONS

Certifications are beneficial in that they give a person credibility in a particular area, can show one's commitment, create a differentiation from other job candidates, and increase earning potential. In addition, many positions in government and corporations require certification because they need qualified personnel with the knowledge and skill set to design, implement, monitor, and protect their critical assets.

One major challenge is determining which certification to earn. Sorting through the vast amount of certifications can be just as overwhelming as the job search. Consequently, before choosing a certification to acquire, it would make sense to do a job search to see what type of certification is required for the positions that are of interest to you. It also helps to determine which certifications are the most wanted by employers. Once you figure out a few certifications that are applicable or required for the jobs of interest, you need to figure out which certification fits your needs. For example, some certifications require "recertification," some require work experience, and all vary in terms of training for the exams. Interestingly, in many of those job postings on LinkedIn and Indeed.com, certifications are listed together, for example: "CISSP [certified information systems security professional] and/ or CISA [certified information systems auditor], CISM [certified information security manager], GIAC [global information assurance certification]." Some of the knowledge domains

A 2021 search on indeed.com revealed the following:

**23,582 jobs required the CISSP** (up from 1,249 in 2009),
**13,521 required the CISA** (up from 1,117 in 2009)
**8,608 required the CISM** (up from 288 in 2009) and,
**4,647 required the GIAC**.

required of these certifications overlap and the employers will often substitute one certification for the other. They are not expecting you to have them all.

All of the most prevalent certifications are covered in this section, but not in any particular order. Similar to the job tasks presented earlier, the certifications have also been categorized in two groups: information systems security certifications (some of these cover digital forensics knowledge) and those that are strictly computer forensic certifications.

### 9.3.1 Information Security Certifications

The first set of certifications presented in this section are granted from the International Information Systems Security Certification Consortium or ISC². The ISC is a well-known organization that certifies and credentials information security professionals. It offers several security certifications. The following is a brief description of the certifications. For further details, see the ISC² website (https://www.isc2.org/):

- **Systems security certified practitioner (SSCP):**
  - *Work experience:* The applicant needs 1 year of work experience in one of the following domains: access controls; cryptography; malicious code and activity; monitoring and analysis; networks and communications; risk, response, and recovery; or security operations and administration.
  - *Knowledge domains:*
    - *Access controls:* Define the operations users allowed to perform on the company systems.
    - *Security operations and administration:* Identify information assets and guidelines to ensure information confidentiality, integrity, and availability.
    - *Monitoring and analysis:* Collect information for identifying and responding to security breaches.
    - *Risk, response, and recovery:* Conduct the process in response to an incident.
    - *Cryptography:* Use techniques to ensure the integrity, confidentiality, authenticity, nonrepudiation, and recovery of encrypted information in its original form.
    - *Networks and communications:* Implement measures to operate both private and public communication networks.
    - *Malicious code and activity:* Implement prevention techniques for malicious activity.

- *Examination:* Take an in-classroom or online course before taking the exam.
- *Endorse the certification:* Once the exam is passed, the applicant needs to have the certificate signed by a credentialed professional in good standing who can attest to the applicant's work experience.
- *Maintain the certification:* Recertification is required every 3 years. This can be obtained by earning 60 credits of continuing professional education (CPE), where at least 10 credits must be completed each of the 3 years. There is also a yearly maintenance fee.
- *Concentrations:* None.

- **Certified authorization professional (CAP):**
  - *Work experience:* 2 years of work experience in one or more of the knowledge domains. Applicants should also have the following knowledge or skill set: IT security, information assurance, information risk management, systems administration, information security policy, technical auditing experience within government departments, and NIST documentation.
  - *Knowledge domains:*
    - *Understand the security authorization of information systems:* Evaluate risk across the enterprise by applying a risk management framework (RMF), revising the organizational structure, and assessing the security controls.
    - *Categorize information systems:* The information systems are categorized to develop the security plan, select security controls, and determine risk. It is based on information included, requirements for the information, and impact on the organization if compromised.
    - *Establish the security control baseline:* The specific controls required to protect the system are based on the categorization, risk, and local parameters.
    - *Apply security controls:* Employ and test the specified security controls in the security plan.
    - *Assess security controls:* Determine effectiveness of the controls meeting the security requirements.
    - *Authorize information systems:* Evaluate risks and determine if the system should be authorized to operate.
    - *Monitor security controls:* Perform continuous monitoring according to the strategy specified by the organization.
  - *Examination:* Take an in-classroom or online course prior to the exam.
  - *Endorse the certification:* Once the exam is passed, the applicant needs to have the certificate signed by a credentialed professional in good standing who can attest to the applicant's work experience.
  - *Maintain the certification:* Recertification is required every 3 years and is obtained by earning 60 credits of CPE and at least 10 credits must be completed each of the 3 years. A yearly maintenance fee is also required.
  - *Concentrations:* None

- **Certified secure software life cycle professional:**
  - *Work experience:* 4 years of experience being involved in the software development life cycle in one of the knowledge domains (following) or 3 years of work experience with a related college degree.
  - *Knowledge domains:*
    - *Secure software concepts:* Know and understand the concepts to consider when designing secure software.
    - *Secure software requirements:* Have effective requirement elicitation and understanding security needs from stakeholders.
    - *Secure software design:* Design secure software architecture and conduct threat modeling.
    - *Secure software implementation/coding:* Utilize secure coding practices to mitigate vulnerabilities and review code to ensure there are no errors in the code and security controls.
    - *Secure software testing:* Integrate software testing to test the security functionality and resiliency to an intrusion as well as how well the software can recover from an attack.
    - *Software acceptance:* Determine if software is free from vulnerabilities in all areas (procurement/supply chain, configuration, compliance, licensing, intellectual property, and accreditation).
    - *Software deployment, operations, maintenance, and disposal:* Deal with security issues surrounding the steady-state operations of the software as well as any security issues surrounding the elimination of the software.
  - *Examination:* Prior to the exam, students participate in webcasts, read the textbook, and attend either an online or classroom education program.
  - *Endorse the certification:* Once the exam is passed, the applicant needs to have the certificate signed by a credentialed professional in good standing who can attest to the applicant's work experience.
  - *Maintain the certification:* Recertification is required every 3 years and is obtained by earning 90 credits of CPE, and at least 15 of the credits must be completed each of the 3 years in addition to a required yearly maintenance fee.
  - *Concentrations:* None
- **Certified information systems security professional (CISSP):**
  - *Work experience:* 5 years of experience in information security. Specifically, the experience can be in two or more of the following areas: access control, telecommunications and network security, information security governance and risk management, software development security, cryptography, security architecture and design, operations security, BC and DR planning, compliance, and physical security.
  - *Knowledge domains:*
    - *Access control:* The concepts, methods, and techniques to develop a secure and effective architecture to protect assets against threats

- *Telecommunications and network security:* Network concepts such as structures, transmission methods, transport formats, and security measures to protect networks, facilitate business goals, and protect against threats
- *Information security governance and risk management:* Developing policies, standards, guidelines, and procedures; determining information assets and understanding risk management tools and practices
- *Software development security:* Effective application-based security controls and software development life cycle
- *Cryptography:* Concepts and algorithms to disguise information and ensure its integrity, confidentiality, and authenticity
- *Security architecture and design:* System and enterprise architecture concepts, principles, and benefits as well as principles of security models
- *Operations security:* Control management, resource protection, vulnerability assessment, attack prevention, and IR
- *BC and DR planning:* Creating response and recovery plans as well as conducting restoration activities
- *Legal, regulations, investigations, and compliance:* Investigating incidents by acquiring and analyzing evidence in compliance with the law and regulations
- *Physical (environmental) security:* Addressing the security of the facility both internally and at the perimeter and having layered physical defense and entry points

- *Examination:* Take an in-classroom or online course prior to the exam.
- *Endorse the certification:* Once the exam is passed, the applicant needs to have the certificate signed by a credentialed, professional in good standing who can attest to the applicant's work experience.
- *Maintain the certification:* Recertification is required every 3 years and is obtained by earning 120 credits of CPE education, where at least 20 credits must be completed each of the 3 years. A yearly maintenance fee is also required.
- *Concentrations:*
  - Architecture (CISSP-ISSAP):
    - *Requirements:* CISSP credential and 2 years of experience in systems architecture
    - *Knowledge domains:* Access control systems and methodology, communications and network security, cryptography, security architecture analysis, technology-related BC planning, DR planning, and physical security considerations
  - Engineering (CISSP-ISSEP):
    - *Requirements:* CISSP credential and 2 years of engineering experience
    - *Knowledge domains:* Systems security engineering, certification and accreditation, technical management, and U.S. government information assurance governance

    – Management (CISSP-ISSMP):
      – *Requirements:* CISSP credential and 2 years of professional management experience
- *Knowledge domains:* Security management practices, systems development security, security compliance management, BC planning, DR planning and law, investigations, forensics, and ethics

This next set of certifications is offered by the global information assurance certification (GIAC), which is a SANS (SysAdmin Audit Network Security) Institute affiliate. They offer many certificates in the following categories: security administration, forensics, management audit, software security, legal, and security expert. The following is a brief description of some of the security-based certifications offered by the GIAC. For further information, see the GIAC website (https://www. giac.org/):

- *Security essentials (GSEC):* This certification is earned based on the knowledge of over 60 objectives. The objectives demonstrate the security tasks that are required of a professional in an IT systems hands-on role (e.g., access control, BC planning/DR, firewalls, honeypots, and incident handling fundamentals).
- *Certified incident handler (GCIH):* This certification is earned based on the knowledge of 25 objectives that facilitate the management of security incidents. Specifically, the focus of this certification is on detecting, responding to, and resolving computer security incidents.
- *Certified intrusion analyst (GCIA):* This certification is earned based on the knowledge of configuring, monitoring, and analyzing the network traffic from an intrusion detection system.
- *Penetration tester (GPEN):* This certification is earned based on the knowledge of finding security vulnerabilities in a network.
- *Web application penetration tester (GWAPT):* This certification is earned based on the knowledge of web application exploits and penetration testing methodology.
- *Certified Windows security administrator (GCWN):* This certification is earned based on the knowledge of securing Windows clients and servers.
- *Assessing and auditing wireless networks (GAWN):* This certification is earned based on the knowledge of security mechanisms for wireless networks as well as how to utilize the tools and techniques necessary to evaluate and exploit the vulnerabilities of a wireless network.
- *Information security fundamentals (GISF):* This certification is earned based on the knowledge of ten objectives that focus on understanding threats and risks to the company's information assets as well as the best practices to protect them.
- *Certified enterprise defender (GCED):* This certification extends the knowledge and skills of the GSEC certification such as defensive network infrastructure, packet analysis, penetration testing, incident handling, and malware removal.

- *Exploit researcher and advanced penetration tester (GXPN):* This certification is earned based on the knowledge of 18 objectives that focus on finding vulnerabilities of target networks, systems, and applications.
- *GIAC security expert (GSE):* This certification requires having earned the GSEC, GCIH, and GCIA certifications as well as having "gold" status (a paper accepted into the SANS reading room) in two of them. There are also options to substitute other GIAC certifications (details on their website). In addition to the prerequisites, this certification is earned based on the knowledge of five knowledge domains: IDS and traffic analysis, incident handling, IT SEC, security technologies, and necessary soft skills.

Each GIAC certification requires passing a proctored exam. The exams vary in length and minimum passing score. Unlike the other GIAC certifications presented here, the GSE also requires the candidate to sit for a hands-on lab. Although SANS training is available, none of the certifications require any specific training. The knowledge tested can be acquired via practical experience or by reading information security publications. All GIAC certifications are valid for 4 years. With the exception of the GSE certification, there are many options for recertification, such as taking the current version of the exam or taking related courses and publishing technical research. The GSE can only be maintained by passing the current version of the exam.

The third set of certifications is offered by the Sherwood Applied Business Security Architecture (SABSA) Organization. There are three distinct levels of certification indicating the stages of proficiency from knowledge and understanding of the concepts to demonstrating and applying the subject matter in order to master the profession. The practitioner (level 2) and master (level 3) levels have prerequisites of the preceding levels. The following is a brief description of the certifications; see http://www.sabsa.org for further details:

- *SABSA chartered foundation (SCF) certificate:* The foundation certificate tests competency in two areas: *strategy and planning* and *security service management.* The testing consists of two multiple-choice exams. Training can be acquired from the SABSA textbook or from one of its training courses.
- *SABSA chartered practitioner (SCP) certificate:* This certificate requires the SCF certificate and passing two additional test modules. The topics of the two additional test modules depend on which of the four specialties are chosen:
  1. Risk management and governance certificate: Focus on *information assurance* and *operational risk management.*
  2. Business continuity and crisis management certificate: Focus on *information assurance* and *BC management.*
  3. Security architecture design and development certificate: Focus on *identity and access management architecture* and *network security architecture and design.*

4. Security operations and service management certificate: Focus on *identity and access management architecture* and *intrusion and incident management.*

- *SABSA chartered master certificate:* This certificate requires the SCF, SCP, and three additional test modules. The topics of the two additional test modules depend on which of the four specialties are chosen:

  1. Risk management and governance certificate: Focus on *management skills, measurement and performance management,* and *information security governance*
  2. Business continuity and crisis management certificate: Focus *on management skills, measurement and performance management,* and *crisis management*
  3. Security architecture design and development certificate: Focus on *management skills, cryptographic techniques,* and *application and web security architecture and design*
  4. Security operations and service management certificate: Focus on *management skills, cryptographic techniques,* and *digital forensics and investigations*

The International Council of E-Commerce Consultants (EC-Council) offers another set of certifications. Similar to some of the other organizations, the EC-Council certifies the information security professional as well as the digital forensics professional. The following is a brief description of the information security certifications:

- *Certified ethical hacker:* This certification covers how to determine the weaknesses and vulnerabilities of a computer system using the tools and skills of a malicious hacker—keeping within the law, of course. The certification also includes the knowledge to prevent and correct vulnerabilities. To be eligible for the certification exam, one can either attend official training or show at least 2 years of information security experience. The website provides an outline of the specific knowledge that will be tested as well as its weight on the exam. The certification is valid for 3 years and can be renewed with EC-Council continuing education credits.
- *EC-Council certified security analyst:* This is essentially an advanced ethical hacking certification. There is a deeper focus on the analytical phase of ethical hacking, such as being able to analyze the outcome of hacking tools and technologies. With these skills one can perform the assessments required to identify and mitigate security risks. To earn this certification, one needs to pass a 50-question exam.
- *Licensed penetration testing:* This certification covers the process of testing the network perimeter defense mechanisms. There are no exams. The candidate needs to have achieved the certified ethical hacker and the EC-Council certified security analyst certifications, as well as be in good standing and not have any criminal convictions.

- *EC-Council network security administrator:* This certification tests the skills to analyze the internal and external security threats against the network. Domain knowledge for this certification include: evaluating the network, evaluating the Internet security issues and design, implementing security policies and firewall strategies, and evaluating vulnerabilities and being able to defend against them.

The Information Systems Audit and Control Association offers the final set of information security certifications covered in this book. It offers four certifications with the knowledge/task domains outlined next. For further details and information, see their website, http://www.isaca.org:

- **Certified information security manager (CISM):**
  - *Information security governance:* Being able to develop and maintain an information security governance framework to ensure that the security strategy of the organization meets the organization's goals and objectives, as well as managing risk and program resources
  - *Information risk management and compliance:* Managing risk in order to meet business and compliance requirements of the organization
  - *Information security program development and management:* Developing and maintaining an information security program that aligns with the organization's security strategy
  - *Information security incident management:* Assuring the ability to plan, establish, and manage the detection of, investigation of, and response to information security incidents.
- **Certified information systems auditor (CISA):** The focus of this certification is based on job tasks and practice. The tasks and knowledge needed for each of the following domains is detailed on their website:
  - *The process of auditing information systems:* Protecting and controlling information systems with IT audit practices
  - *Governance and management of IT:* Ensuring that leadership, organizational structure, and processes are able to achieve the organization's objectives and strategies
  - *Information systems acquisition, development, and implementation:* Meeting the organization's strategies and objectives when acquiring, developing, testing, and implementing the information systems
  - *Information systems operation maintenance and support:* Meeting the organization's strategies and objectives when operating, maintaining, and supporting the information systems
  - *Protection of information assets:* Ensuring that the organization's security policies, standards, and procedures protect the confidentiality, integrity, and availability of the information assets
- **Certified in the governance of enterprise IT:**
  - *IT governance framework:* Design, develop, and maintain an IT governance framework that includes leadership and organizational structures and processes. The framework must align with the company's

governance, implement good practices to control the information and technology, and ensure compliance.

- *Strategic alignment:* Ensure that IT supports the following: integrating IT strategic plans with business strategic plans and continuing to achieve business objectives and optimizing business processes by aligning IT services with business operations.
- *Value delivery:* Ensure that IT and the business fulfill their value management responsibilities (IT services and assets contribute to the value of the business).
- *Risk management:* Ensure that the framework exists to manage and monitor IT business risks.
- *Resource management:* Ensure that the competence and capabilities of IT resources can meet the demands of the business.
- *Performance measurement:* Set, monitor, and evaluate business-supporting IT goals.
- **Certified in risk and information systems control:**
  - *Risk identification, assessment, and evaluation:* Execute the enterprise risk management strategy.
  - *Risk response:* Develop and implement risk responses to address risk factors and incidents.
  - *Risk monitoring:* Communicate risk indicators to stakeholders.
  - *Information systems control design and implementation:* Design and implement information system controls.
  - *IS control monitoring and maintenance:* Effectively monitor and maintain the information systems controls.

## 9.3.2 DIGITAL FORENSIC AND FORENSIC SOFTWARE CERTIFICATIONS

Like the information security certifications, there are many digital forensic certifications. Some are offered by product vendors, some by professional organizations, and some by training providers. Until there are more degrees offered in this field, it would be safe to say that with this highly specialized profession, a certification is most likely a must. The following are some of the vendor-neutral certifications.

### 9.3.2.1 Digital Forensic Certifications

In the previous section we covered the security-based certifications offered by the GIAC. Next we will present a brief description of some of the forensics-based certifications offered by the GIAC (https://www.giac.org/) as well as a few others:

- *Certified forensic analyst:* Offered by GIAC, the focus of this certification is on the skills and knowledge necessary to collect and analyze data from digital media. One can prepare for the proctored exam by taking the SANS training course, Advanced Computer Forensic Analysis and Incident Response. This certification must be renewed every 4 years by either taking the exam again or earning 36 certification maintenance units (CMUs). The

CMUs range from attending more training to publishing a related paper or book.

- *Reverse-engineering malware:* Offered by GIAC, this certification tests the knowledge and skills one needs to protect an organization from malicious code by reverse-engineering malware. One can prepare for the proctored exam by taking the SANS training course, Reverse-Engineering Malware: Malware Analysis Tools and Techniques. This certification must be renewed every 4 years by either taking the exam again or earning 36 CMUs. The CMUs range from attending more training to publishing a related paper or book.

- *Certified forensic examiner (GCFE):* Offered by GIAC, the candidate with the GCFE will be able to conduct investigations that include e-discovery, forensic analysis and reporting, evidence acquisition, browser forensics, and tracing user and application activities. One can prepare for the proctored exam by taking the SANS training course, Computer Forensic Investigations—Windows In-Depth. This certification must be renewed every 4 years by either taking the exam again or earning 36 CMUs. The CMUs range from attending more training to publishing a related paper or book.

- *Certified computer examiner (CCE):* The next certification is the CCE, which is the primary certification offered by the International Society of Forensic Computer Examiners. The International Society of Forensic Computer Examiners is a private organization that offers training and certification. The CCE also performs research and development into new technologies and methods for computer forensics (http://isfce.com/). This certification includes a practical exam as well as an online written exam. A candidate who passes the online exam will have 90 days to complete the practical exam. Recertification consists of 40 CPE credits and practice in the field, which needs to occur after 2 years.

- *Certified forensic computer examiner (CFCE):* The International Association of Computer Investigative Specialists, a nonprofit organization that trains computer forensics professionals, offers a 2-week course designed to prepare candidates for the CFCE. The training course is not required to enter into the CFCE certification process. The CFCE process consists of completing a series of practical exercises. Upon successful completion of these exercises, the candidate needs to pass a comprehensive written exam. The knowledge domains covered are as follows: pre-examination procedures and legal issues, media examination and analysis, data recovery, specific analysis of recovered data, reporting and exhibits, and defense and presentation of findings. The CFCE requires a recertification every 3 years that includes an exam and 60 hours of continuing education in the field. For further details, see http://www.iacis.com.

- *Computer hacking forensic investigator:* As mentioned in the previous section, the EC-Council also offers a computer forensic certification. This certification focuses on the process of detecting hacking attacks and properly extracting evidence to report crimes, conduct audits, and prevent future attacks. The candidate is awarded the certification upon successful completion of the 150-question exam. Further details can be found on the EC-Council website (https://cert.eccouncil.org/).

- *Professional certified investigator (PCI):* ASIS International offers the professional certified investigator. This certification is awarded upon successful completion of a multiple-choice exam covering case management, investigative techniques and procedures, and case presentation. The PCI has a 3-year recertification term that can be accomplished with 45 CPE credits.
- *Certified computer forensics examiner (CCFE):* The Information Assurance Certification Review Board (IACRB) offers the CCFE. This certification tests a candidate's knowledge of the following knowledge domains: law and ethics, investigation process, computer forensic tools, device and evidence recovery and integrity, file system forensics, evidence analysis and correlation, evidence recovery, and report writing. The CCFE requires a 50-question multiple-choice exam and a practical exam (given upon successful completion of the multiple-choice exam). The IACRB is a professional nonprofit organization whose sole mission is to certify individuals. One can get training for the exam from an IACRB-approved training provider. See http://www.iacertification.org/ for further information.

### 9.3.2.2 Forensic Software Certifications

In addition to the vendor-neutral certifications, two well-known forensic software packages offer certifications: EnCase (ENCE) and AccessData Certified Examiner (ACE). These certifications appear in many job postings (along with other certifications, of course); therefore, it was added to this listing of certifications.

Guidance Software's EnCase is a one of the industry-standard computer investigation tools. It essentially enables the investigator to acquire the data and evidence from many devices to create reports and maintain the integrity of the evidence. The **EnCE** is the EnCase certification offered by Guidance Software that ensures the mastery of the computer investigation methodology as well as effective use of the EnCase software during an investigation. This certification requires 64 hours of computer forensic training or 1 year of computer forensic work experience, the approval of the EnCE application, passing the written exam, and passing the practical exam. The candidate has 60 days to complete the practical exam. The certification renewal has a few options, but essentially requires 32 hours of training every 3 years. For further information, see www.guidancesoftware.com.

Another industry-standard computer forensics software package is AccessData Group's **Forensic Toolkit (FTK)**. FTK is similar to EnCase in that it is used in investigations to image, analyze, and report. The ACE demonstrates proficiency with FTK. The certification consists of one multiple-choice exam. The candidates need to have access to a computer with a licensed version of FTK, since the questions relate to analyzing an image in FTK by the candidate. One year after passing the ACE exam, the candidate is required to pass an online practical examination. After the practical exam, the certification needs to be renewed every 2 years. There are free online videos to prepare for the exam. For further information, see www. accessdata.com.

## COMPUTER ANALYSIS RESPONSE TEAM (CART)

If you have any aspirations to work for the U.S. government, there are opportunities at the many different government agencies as most have a computer forensic need. The four main FBI priorities that require computer forensic support are *innocent images* (a program that identifies pedophiles who use the Internet to lure children or spread child pornography), *counterintelligence* (combating the infiltration of the U.S. intelligence community by foreign intelligence services), *counterterrorism* (addresses terrorist threats), and *criminal* (white collar crimes such as fraud and public corruption). Each FBI field office has a computer forensic unit called the CART whose primary function is to retrieve evidence from electronic media. CART training, available for both agents and professional support within the Bureau, includes some of the industry certification exams discussed in this chapter. For more information, see www.fbi.gov.

## REFERENCES

BlueFin. September 9, 2021. The biggest data breaches of 2021 so far. https://www.bluefin. com/bluefin-news/2021-biggest-data-breaches-so-far/ (retrieved November 22, 2021).

Cichonski, P., Millar, T., Grance, T. and Scarfone, K. August, 2012. Computer security incident handling guide. NIST special publication 800-61, revision 2.

Eckelberry, A., Dardick, G., Folkerts, J., Shipp, A., Sites, E., Steward, J. and Stuart, R. 2007. Technical review of the trial testimony State of *Connecticut v. Julie Amero*. http://www.sunbelt-software.com/ihs/alex/julieamerosummary.pdf (retrieved May 18, 2012).

Liu, F., Tong, J., Mao, J., Bohn, R., Messina, J., Badger, L. and Leaf, D. 2007. NIST cloud computing reference architecture: Recommendations of the National Institute of Standards and Technology. NIST publication 500-292. http://www.nist.gov/customcf/get_pdf.cfm?pub_id=909505 (retrieved July 18, 2012).

SANS. n.d. InfoSec acceptable use policy. http://www.sans.org/security-resources/policies/Acceptable_Use_Policy.pdf (retrieved July 18, 2012).

Scholl, M., Stine, K., Hash, J., Bowen, P., Johnson, A., Smith, C. D. and Steinberg, D. 2008. An introductory resource guide for implementing the Health Insurance Portability and Accountability Act (HIPAA) security rule. NIST special publication 800-66. http://csrc.nist.gov/publications/nistpubs/800-66-Rev1/SP-800-66-Revision1.pdf (retrieved July 18, 2012).

Schwartz, M. June 25, 2012. Senators float national data breach law, take four. *Information* Week.

## NOTES

1  A pop-up is a browser window that appears out of nowhere when a web page is visited. Sometimes the pop-ups are advertisements and sometimes they are malicious programs that will install the undesirable content to the machine upon clicking.

2  The technical details of this case can be found in Eckelberry et al. (2007).

3  Computer forensics is also known as digital forensics. The terms are used interchangeably in this book.

4  A warrant is an order from a legal authority that allows an action such as a search, an arrest, or the seizure of property.

# 10 Theory to Practice

> Many of life's failures are people who did not realize how close they were to success when they gave up.
>
> *—Thomas Edison*

## 10.1 INTRODUCTION

It is time to put the concepts you have been reading about into practice. The previous chapters gave you the foundation needed to understand the implications of decisions made during real security incidents. There are three cases presented in this chapter. I will present the first two cases in a story format as they happened. The third case recounts the details of a high-profile litigation. After each case, an *after-action report* (aka a postmortem) will be presented. A postmortem is an exercise that a team performs to review a project with the goal that improvements will be made for the next project. Organizations that want to learn from their mistakes and continuously improve do this type of exercise after a challenging project. It is important to note that this is also a great exercise to do when a project is successful so that things that worked well are repeated. Since the presentation in this chapter is for illustrative purposes, the postmortems have been pared down a bit. If you are interested in learning the complete process to perform a postmortem, read "An Approach to Postmorta, Postparta and Post Project Reviews" by Norman Kerth (2013).

## 10.2 CASE STUDY 1: IT IS ALL FUN AND GAMES UNTIL SOMETHING GETS DELETED[1]

You are the new chief information security officer (CISO) of a large organization. As with most senior-level executives, you are responsible for maintaining and overseeing many moving pieces in the organization; however, since you have only been in place a few days, you are still getting a grasp on the organization as a whole. You are still determining the critical assets, the topology of the network, and, most important, the personnel resources you have to help you do your job of securing the infrastructure of this company. Unfortunately, you have already noticed that industry best practices in the area of security have not found their way to this organization ... yet.

You begin your day like any other day with meetings and e-mail. It is about 11:00 a.m., and you decide to take a break. You stop in the restroom and overhear a conversation between an engineer and the IT administrator who walked in after you; they are discussing a recent incident. Here is what you hear:

*Paul (engineer):* Hey, David, how's it going?

*David (IT administrator):* Pretty good, Paul. Sorry I didn't meet you guys last night for happy hour. I noticed at the end of the day that the web server went offline.

*Paul:* Oh, did you get it back up and running?

*David:* Yeah, it was no big deal. The global.asa[2] file was deleted. We got it back up and running in no time.

*Paul:* That's cool. I'll let you know when we plan another happy hour.

> *You are alarmed at the casual conversation regarding this incident. You leave the restroom and wait for David in the hall, who realizes you overheard the conversation when he sees you:*

*CISO:* Hey, David. I overheard your conversation about the web server issue. I would like to review the incident report.

*David (now looks like a deer caught in the headlights):* I didn't fill out a report because it was an easy fix.

*CISO:* Let's discuss this in my office.

Before you give David a lecture on why documenting an incident and following an incident response process is crucial, you decide to listen to the facts first. You begin your discussion in line with the *incident response* process:

*CISO:* David, please first explain in detail how you detected this incident.

*David:* Sure. I uploaded a new PDF to the database and wanted to test how the information from the PDF was displayed on our website, but I couldn't access the website because it was offline. I checked the root directory of the server and noticed the global.asa file was missing. I looked in the log file and determined the global.asa file was deleted yesterday, which in effect "breaks" the website. When I looked around a little more on the server, I noticed that a lot of game and movie files had been uploaded. I figured someone just uploaded them to play a game or watch a movie because I did an antivirus scan and didn't see any evidence of someone remotely controlling the server.[3] Before I did anything else, I called Tim on the Network Security Team to keep him informed. He suggested searching for more malware in the form of spyware and Trojan infections. When nothing was discovered, the network security team declared this was not a hack, and thus not an incident. So, I started recovering the system by deleting the movies and games and uploading the backup of the global.asa file.

At this point, David feels confident about his explanation because he followed the correct process by calling the Network Security Team and performing the actions prescribed by the team. The only thing he is a little worried about is whether the CISO is going to question why the pirated movies were not reported to law enforcement, so he decides to clarify before the CISO says anything: "We didn't report the incident regarding the movies because we couldn't afford to have our server taken offline. Law enforcement would have needed to review the server for evidence, right?"

You are silent for a few minutes. There is so much to do here, so you ignore his last statement for now and think about their process. You are thinking that David did go through the IR process: he *detected* the incident, *contained* and *eradicated* the incident by deleting the games and movies, and *recovered* the web server by restoring the missing file. However, the analysis was clearly lacking in thoroughness, which caused a premature declaration that this was not a hack. You break the silence with the million-dollar question: "You didn't determine how the games and movies were uploaded. Did you check the server logs? If the vulnerability is still there, you didn't solve the problem!" David starts to sweat. He is really not to blame; he was following the advice of the security team. However, before David can answer, you say, "David, please get Tim and let's meet in the conference room in ten minutes."

You go straight to the conference room and wait. David and Tim arrive and sit down. You start the conversation:

*CISO:* Tim, David has brought me up to date with this incident. I am concerned about the lack of thoroughness in the investigation process being utilized, but we can work on that after we resolve the current issue. First, we need to determine the vulnerability that allowed the uploads so that it can be eliminated. What is the status of the last vulnerability scan?

*Tim:* I am not sure when the last scan was performed. I will get back to you after I look at the logs that we do have from the intrusion detection system and check some of the access control mechanisms.

*CISO:* David, I want you to document what has occurred so far and also work with Tim to document the rest of the investigation. Please keep me informed.

After 3 hours go by, David appears in your doorway and says:

I have an update. Tim used a forensic tool to evaluate the server over the wire. With this tool, he was able to connect to the server and *collect* the information needed for analysis in a proper forensic manner. He determined that the vulnerability was that the FTP account was configured without a password. When he *analyzed* the data in the log files, he found that malware was in fact introduced to the web server. However, the games and movies were not uploaded onto the server using the open FTP account. They were uploaded via a back door—the second vulnerability. We suspect now that our server was being utilized as a peer-to-peer (P2P[4]) mechanism.

You contain your urge to gloat and say,

So, a rootkit[5] was introduced through the FTP account and the games and movies were introduced though a back door access created by the rootkit. If I am correct, the games will be reinstalled shortly since the recovery performed yesterday did not include elimination of these vulnerabilities.

David looks a little deflated because you stole his punch line. David continues, "Yes, the games and movies were in fact reinstalled after I deleted them last night, so you are correct."

David gets a call from Tim:

*David:* Hi, Tim. Yes, I am in his office.

David puts the phone on speaker and Tim says,

> Hi, everyone, here is the update: Upon analysis of the log files, we determined that the hacker deleted the global.asa file yesterday. It appears that this hacker was cleaning up unneeded files on the server, probably to make room for additional movies and games so that he wouldn't get noticed. During that process, he accidentally deleted the global.asa file. It also appears that, according to the log file, there were 500 individual downloads last week alone. So, I just deleted the rootkit and secured the FTP account, which should mitigate any further issues. The worst part of this incident is that, according to the log files, our server has been compromised for a year.

You tell them both that they did a great job, but they are not finished. Tim now needs to pass the evidence of this intrusion to law enforcement for further analysis as they may be able to determine the identity of the criminal. Tomorrow, you will break the news about ramping up their incident response process as well as explain how this incident could have easily been avoided.

### 10.2.1 After-Action Report

#### 10.2.1.1 What Worked Well?

The most obvious thing this organization did well was to hire a CISO, since it is clear it needed a leader to implement security best practices into its operations. For example, the company had the right tools, but did not use them effectively. There was intrusion detection, but the parameters were set so liberally that hardly any events were logged. A vulnerability scanner had been installed, but it was not configured properly. Before the CISO came onboard, this organization was getting hit with virus after virus, so the CISO was tasked to create a security posture for the company.

#### 10.2.1.2 Lessons Learned

Even though you have the right tools, training is needed to use them effectively. For example, although the company had a vulnerability scanner, a daily or even weekly vulnerability scan would have showed the open FTP account early on. Not everyone needs a daily vulnerability scan. The asset value will determine the frequency and thoroughness of your scanning because each scan requires resources. Someone has to review the vulnerability report!

Have an incident response team in place. The network security team's first priority is not incident response; therefore, they were not prepared to investigate this incident. It was clear that their goal was to get the server back up and running. The incident response team will also make sure the organization is prepared for an incident and has a process in place to handle one.

#### 10.2.1.3 What to Do Differently Next Time

These can also be called the after-action items:

1. Make sure the intrusion detection system has current signature files. Signature files will help the system recognize known malicious threats. This is similar to the way in which antivirus applications detect malware.

2. Migrate into an enterprise server format where the technical controls would be more rigorous. In other words, the company needs a centralized server resource as opposed to having each department run its own servers. This will help the company analyze and secure the servers in a consistent manner. In addition, a company should hire people to manage that effort.
3. Implement incident response training for all of the IT administrators. This will help them recognize incidents as well as understand the importance of ensuring that the incident response process is followed.
4. Review and update the change management request process to ensure that proper access control is implemented.
5. Conduct regular vulnerability scans.

The 2012 LinkedIn database breach where hackers obtained millions upon millions of access credentials was a wake-up call to companies that have not kept a close enough eye on their organization's security plan. Here are nine techniques that a CISO can employ to improve the effectiveness of an organization's security posture (Schwartz 2012):

1. *Deploy CISOs in advance:* This is part of being prepared. Would you move to a town that did not have a fire department, police department, or hospital? Hire the CISO before the security breach happens—not after.
2. *Acknowledge how CISOs reduce security costs:* According to the Ponemon Institute (2012), the cost of data breach attacks has declined from $7.2 million to $5.5 million. In addition, they reported that the organizations that employed CISOs had an $80 cost savings per compromised record. Companies that outsourced this function only saved $41 per compromised record. The reason a CISO reduces costs is that he or she can help facilitate security best practices that have been proven successful.
3. *Allow CISOs to help guide new technology decisions:* The evolution of technology is ongoing. The CISO needs to be accepting of new technologies in order to factor them into the organization's overall security profile.
4. *Make CEOs demand security posture details:* Effective communication between the CEO and CISO is a must. In other words, the CEO needs to have an appreciation and understanding of the organization's security posture just as he or she has an appreciation and understanding of the organization's current sales.
5. *Treat information security as a risk:* Something as simple as a phishing attack on a company can compromise the security of critical information. The CISO needs to be well informed of all vulnerabilities in the organization as well as vulnerabilities at organizations that share any of his or her organization's computing resources.
6. *Consider a placeholder CISO:* If your company does not have a CISO, consider outsourcing the position to a reputable security company until the needs of the organization are determined.
7. *Identify crown jewels:* In part of the risk analysis, determine the value of the critical assets. In addition, risk should be reassessed periodically. For

example, if a password file has doubled the number of users, increasing its protection should be a priority.

8. *Beware of a false sense of security:* Use a third party, who may see things that you do not, to assess the risk and security posture of the organization.

9. *Treat advanced threats as common:* Consider advanced persistent threats (APTs, discussed earlier) as more prevalent than ever. The standard information security defense should never be standard; it needs to evolve as the threats evolve.

## 10.3 CASE STUDY 2: HOW IS THIS WORKING FOR YOU?

Let us fast-forward 2 years at the same organization and see how well the CISO's security plan has worked out. Over the 2 years, a few people have been hired for the computer incident response team (CIRT), so we have a new cast of characters in our story. We have Jenny leading the CIRT team and Alex and Justin working with her. We also still have David, who continues in his IT administrator role but has since been trained in incident response per one of the after-action items in our last case study.

Another one of the changes the CISO instated was to write and enforce an acceptable use policy (AUP). We discussed AUPs in Chapter 4. One of the restrictions at this organization, according to the AUP, was that instant messaging (IM) is not allowed. It was felt that IM was a distraction to the employee and, more important, it was deemed a security risk to the network. IM tools are security risks because they can circumvent the security measures (e.g., employee can casually send out confidential information) of the organization as well as become a conduit for worms[6] and viruses.

The exact AUP excerpt read as follows:

INSTANT MESSAGING and CHAT Services:
    Use of instant messaging or chat applications on the company network is not acceptable.

To monitor the IM restriction, the intrusion detection system was configured to generate an alert if IM was being utilized. Well, one afternoon, Alex informed Jenny that the IM alarm had been triggered. Upon analysis, the identity of the employee was discovered, but the CIRT needed to analyze the hard drive to determine the nature of the messages. If the messages were inappropriate, this would be grounds for the employee's dismissal. Alex informed Jenny of the situation and they made a plan. They needed to inform the director of Human Resources (HR) of the violation and plan a time to approach the employee to confiscate the employee's hard drive. The team decided to approach the employee within the hour.

The CIRT team and director of HR gathered and approached the employee. Jenny said, "It has come to our attention that you are in violation of this organization's acceptable use policy. We will need to confiscate your hard drive." The employee obviously was stunned, but was cooperative.

The CIRT team brought the hard drive back to their lab, made a forensically sound[7] copy of the hard drive, and began their analysis. They knew the instant massager client, so they needed to analyze the application to determine the messages that were received and sent. They discovered that the messages were of an inappropriate nature, which is grounds for dismissal. They needed to follow appropriate procedures to store the evidence in the event that the employee decided to contest the dismissal. For now, the situation was handed over to Human Resources to dismiss the employee.

### 10.3.1 After-Action Report

#### 10.3.1.1 What Worked Well?

The CISO has done an excellent job implementing industry's best practices into this organization's security profile. An AUP policy was written, implemented, and followed. The organization had a dedicated team to respond to the incident and they followed forensically sound procedures to analyze the situation.

#### 10.3.1.2 Lessons Learned

The importance of implementing industry best practices both to secure the company assets and to be able to respond to incidents is priceless.

#### 10.3.1.3 What to Do Differently Next Time

Nothing! Well done!

## 10.4 CASE STUDY 3: THE WEAKEST LINK[8]

### 10.4.1 Background

Roger Duronio was dissatisfied with his yearly bonus from his employer, the financial services company, UBS-Painewebber (UBS-PW). Like many companies, after

**TO OUTSOURCE OR NOT**

Some companies choose to outsource the security function to a third party because they can save money or the third party can do a better job for the same money. Examples of outsourced functions could be hiring consultants to help deal with a data breach or hiring them to store your data.

Outsourcing needs to be thought about carefully because your company is ultimately responsible in the case of a security breach. Conducting a risk assessment to help with that decision is necessary (Condon 2007): Determine the potential impact on the organization if a data breach occurs and determine if the outsourcing company will make your data vulnerable. According to the Ponemon Institute (2012), 41% of organizations had a data breach caused by a third party (outsourcers having access to protected data, cloud providers, and/or business partners). Most likely, determining the quality of service you will get from the security firm will be confirmed by references from other customers and a site visit (Burson 2010).

the events of 9/11, profits were down at UBS-PW, which affected the employee bonus program. On February 22, 2002, the bonuses were distributed. Duronio's bonus was $15,000 less (his compensation for the year would be $160,000 instead of $175,000) than what he expected, even though the employees were informed previously that this bonus reduction would be happening. Duronio had a history of being dissatisfied with his pay. The prior year he had approached his boss for a raise. His boss was able to approve a $10,000 salary raise; however, the boss felt Duronio was still unsatisfied with his compensation. This was apparent when Duronio received his bonus on February 22, 2002. After receiving his bonus, he went straight to his boss and demanded the remainder be awarded. Otherwise, he would quit that very day. The boss made an attempt to have the full bonus awarded, but was not successful. When he went back to give Duronio the bad news, his boxes were already packed. His vengeful plan was already in the works.

Duronio's revenge on UBS-PW caused him to be charged with *securities fraud* (count 1), *mail fraud* (counts 2 and 3), and *fraud and related activity in connection with computers* (count 4). In the high-profile case, the U.S. Department of Justice hired computer forensics expert Keith Jones to testify on behalf of the prosecution. The defense hired Kevin Faulkner as their forensics expert.

### 10.4.2 THE CRIME

On Monday, March 4, 2002, Duronio, a former systems administrator for UBS-PW, executed a logic bomb within its network that disabled nearly 2,000 of the company's servers. He planted the logic bomb prior to his exit from the company. A logic bomb is malicious code inserted into an application that will execute when the specified condition is met. His logic bomb was set to execute when the stock market opened at 9:30 a.m. EST on March 4, 2002. The code had four components:

1. *Destruction:* The server would delete all files.
2. *Distribution:* The bomb would be pushed from the central server to 370 branch offices.
3. *Persistence:* The bomb would continue to run regardless of a reboot or power down.
4. *Backup trigger:* If the logic bomb code was discovered, another code bomb would execute the destruction.

The logic bomb was only the first part of the plan. The second part of his plan was to profit from this attack. Duronio purchased 330 PUT[9] options ($25,000 worth) of UBS-PW shares. He was essentially betting on the fact that he would make money when the stock lost value due to his logic bomb attack. UBS-PW reported a $3 million loss[10] in recovery from this attack.

### 10.4.3 THE TRIAL

Mr. Jones, the forensics expert for the prosecution, had his work cut out for him. He had to piece together the puzzle that proved the deceptive actions of Duronio as well as present the facts of the case in a way that could be understood by the jury. The forensic expert for the defense, Kevin Faulkner, had to prove the opposite. The trial went on for 5 weeks.

### 10.4.3.1   The Defense

The goal of the defense was to show that evidence presented by the prosecution was incomplete and unreliable. Their main focus was on the fact that there was no mirror image of the data and consequently no way to prove that Duronio was the attacker. In reference to the fact that there were only backup tapes of the hard disk files to analyze because a forensic image (a bit-for-bit copy) of the drive was not taken, Mr. Faulker said, "I couldn't look at all of the data." He stated, "To preserve digital evidence, a forensic image is best practice." He only had 6.5 gigabytes of data from a 30-gigabyte capacity server to analyze. The defense attorney questioning Faulkner attempted to assert that a forensic analysis of backup tapes is not sufficient to make any hard conclusions. In addition, the attorney was putting into question the chain-of-custody of the data because the backup tapes were handled by another forensics company no longer involved in the case. This former forensics company also had a reputation of hiring hackers which, in their opinion, put the integrity of the forensics company as well as the integrity of the data previously handled by hackers into question.

The defense attorney also questioned Mr. Jones about the validity of the analysis using only backup tapes of hard disk files instead of a bit-for-bit copy of the servers. Mr. Jones testified that taking an image of damaged servers would not have aided in the success of the analysis. He felt the amount of data available was sufficient to draw conclusions.

### 10.4.3.2   The Prosecution

Over 5 days, Mr. Jones testified that Duronio's actions caused the UBS-PW stock trading servers to be inoperable. He was able to extract IP address, date, and time information that connected the attacker to the specific servers and confirmed when and where the attacker had planted the logic bomb. The IP address pointed directly to Duronio's home in all cases but one. The exception pointed to Duronio's workstation at UBS-PW. The U.S. Secret Service also found parts of the logic bomb code on two machines within Duronio's home in addition to a hard copy printout of the code.

Mr. Faulkner pointed out the alleged holes in the prosecution's testimony. He testified that the log data in general are poor forensic evidence. The logs that were used by the prosecution were the VPN, WTMP, and SU (switch user logs show when users switch to *root* user[11] access). It is important to note here that root user access, which Duronio had, would be necessary to plant a logic bomb. Mr. Faulkner also provided a few other facts that attempted to put the attacker ID into question:

1. The log data are not reliable, as they can be edited by the root user.
2. The log files data would not be able to identify whether someone accessed the server using a back door.[12]
3. There was, in fact, back door entry to the server in question.

4. Although the time of their access was not identified to match the time of the logic bomb insert, two people (only identified via login ID) accessed the server using the back door.
5. There were two other current systems administrators who were also employed at UBS at the time of the attack who could have been the attacker. However, the two other system administrators were cleared of any suspicion of direct involvement after the first forensic investigation team (no longer working on the case) analyzed their machines. That company did find a few strings of the logic bomb code in the swap space[13] on one of the systems administrator's machines. But there was no other criminal evidence found on that machine. They also did not find any other information to show that the code bomb existed on that machine. Interestingly, the data from those two machines were destroyed when the first forensic company (recall the chain-of-custody issue mentioned earlier) was bought out by another company.

The testimony of Mr. Jones clarified that the data analyzed pointed to the user with the ID of "rduronio." The log data showed that this user was accessing the server from inside Duronio's home. Mr. Jones also clarified that the reason backup tapes were used instead of a bit-for-bit copy of the data was that the server data were damaged—so an image would not have been helpful. In addition, the IT workers at UBS were focusing on getting the system back online at the time of the attack, so the recovery efforts would have written over data left on the server. Mr. Jones felt strongly that anything additional from a bit-for-bit copy would not contradict what was already discovered on the backup tapes anyway.

During the redirect[14] questioning, Mr. Faulkner was asked by the defense attorney, "Do you have a bottom line as to which username is responsible for the logic bomb?"

Mr. Faulker replied, "Root."

Since there were other system administrators with root access, the defense attorney asked a follow-up question, "Is there evidence which username, acting as root, was responsible?"

Mr. Faulker replied, "No."

Assistant U.S. Attorney Mauro Wolf asked one additional question that turned it all around, "Bottom line…root did it. Roger Duronio could have acted as root?"

Mr. Faulker replied, "Yes."

### 10.4.3.3   Other Strategies to Win the Case

1. *Defense:* It was a conspiracy against Roger Duronio.

   a. The U.S. Secret Service must have planted the evidence in Duronio's home. First, there was an unknown fingerprint on the hard copy of the code found in the house. Second, the Secret Service removed the computers from the house before the forensic image was taken of the machine. This may have been the reason they discovered the logic bomb code on the computers back in their office instead of in Duronio's home—because they put it there!

b. The expert witness for the prosecution was biased and had an agenda because he was part owner of the company hired to do the forensic analysis.

c. UBS was hiding evidence. The data from the workstations of the other two systems administrators were destroyed. The first forensics company was bought out and the evidence was destroyed in the process; this was not the doing of UBS. In addition, recall that the first forensics company hired hackers; therefore, the evidence they touched must be polluted.

d. At one point, the defense also attempted to blame a scheduled penetration test of their system by Cisco.

2. *Prosecution:* Not much is needed here as they already had discovered enough data to convict Duronio. So, they pointed out that the background of the defense's forensic examiner was weak.

a. He had 2.5 years of forensics experience, most of which was gained during this case.

b. The defense's forensic examiner did not come to any conclusions following his forensic analysis.

c. The theories of the defense were all red herrings.[15] Why would all of those people (UBS, U.S. Secret Service, Cisco, and the first forensics company) be after Roger Duronio?!

### 10.4.3.4 Verdict

Roger Duronio was found guilty. He was sentenced to 97 months without parole. He was also ordered to make $3.1 million in restitution to UBS Pain Webber.

### 10.4.4 AFTER-ACTION REPORT[16]

#### 10.4.4.1 What Worked Well for UBS-PW?

1. *Resources:* The UBS IT executives had a plan and were able to get the system back up and running with the help of hundreds of consultants from IBM as well as hundreds of people from their own staff.

2. *Look for outside help:* They used a third party to lead the recovery effort (IBM) as well as a third party to do the investigation. Outsiders take an objective view of the problem. This is critical when an insider is suspected to be the cause of the problem.

3. *Find the problem and go nonstop:* The dedicated staff that worked nonstop on the problem was very effective in addressing the issue. They also did not stop until the problem was eradicated and the system was recovered.

4. *Backup:* The backup tapes restored the servers that were damaged.

5. *Learn from the experience:* UBS-PW did a postmortem on the event to learn from the experience.

#### 10.4.4.2 What to Do Differently Next Time

1. *Remember that humans are the weakest link:* From weak passwords to disgruntled employees with access to critical systems—do not discount the damage that can be done.

2. *Enhance log reports:* The logs were good but could have been better. For example, they showed who switched to the root but not which commands the root ran on the system.
3. *Limit root privileges:* Systems administrators should have root privileges only necessary to do their jobs. They do not need access to the whole system.
4. *Break the trust relationship:* Use better authentication between branch servers. In this situation, no authentication was required, so the logic bomb was easily pushed out to each server from the central server.
5. *Use encrypted protocols:* Use secure sockets layer (SSL) when allowing remote access to computers.

## 10.5   CASE STUDY 4: DEALING WITH PHISHING USING THE OODA LOOP

The 2018 Verizon's Data Breach Investigation Report (DBIR) (https://www.phishingbox. com/downloads/Verizon-Data-Breach-Investigations-Report-DBIR-2018.pdf) states,

> Normally when we start talking phishing, it's all doom and gloom. But you know what? Most people never click phishing emails. That's right, when analyzing results from phishing simulations the data showed that in the normal (median) organization, 78% of people don't click a single phish all year. That's pretty good news. Unfortunately, on average, 4% of people in any given phishing campaign will click it, and the vampire only needs one person to let them in.

The 2018 Verizon DBIR also states, "With 'only' 13% of breaches featuring phishing, it may appear to be feeding from the bottom this year." In addition, banking Trojan botnets were removed from these number; however, 70% of breaches associated with nation-state or state-affiliated actors involved phishing.

Why is phishing relevant to your organization? An end-user falling prey to a phish can not only be an entry point for a ransomware attack but can also be the start of an advanced persistent threat (APT) where bad actors infiltrate your networks and systems for many nefarious and damaging purposes. Understanding the potential financial losses will support a decision to implement security awareness training.

Using the OODA Loop, lets us examine the process to understand and improve your organization's ability to prevent successful phishing attacks. The first step, **observe**, is to collect information (metrics) around security awareness training, metrics such as the scores of awareness tests and percentages of employees completing training. Collect information on previously known incidents of successful phishing in your organization, as well as metrics of blocked phishing attempts.

With the collected information, or the lack thereof, you would now **orient** (e.g., analyze) what you know. Part of the orient phase is to know how the risk, or probable financial impact, fits into your risk appetite (e.g., the level of loss you are willing to accept). It would be a perfectly good business decision (**decide**) to accept the risk, and loop back to observe to monitor the situation for any change (**act**).

Although in this example, phishing can lead to catastrophic loss from ransomware, loss of intellectual property, or a significant data breach. This leads to a new, **tangent loop**. The tangent loop would begin with a collection of information (observations) about the implantation of various controls, including but not limited to the increase of specific phishing-related security awareness training, the implantation of multi-factor authentication (MFA) on critical, or perhaps all logins, e-mail controls, etc. During the orient phase, you would analyze each control for its ability to reduce probable financial impact, then decide if it makes business sense (the cost is far less than the risk for example), then decide to impellent the control or controls.

As your program matures, you continue to use the OODA Loop to monitor the success (or failures) to continuously improve your posture. An excellent framework to use to assist in the understanding of probable financial impact (orient) is FAIR, the Factor Analysis of Information Risk. More details can be found at the FAIR Institute (https://www.fairinstitute.org/).

## 10.6   CASE STUDY 5: DEALING WITH INCIDENT RESPONSE USING THE OODA LOOP

Cyber security incident response is a very effective use of OODA. For example, in the Commonwealth of Pennsylvania, people getting malicious links served up by one of their agency websites started the OODA Loop process:

- While they were **observing** the first sign of an issue, another agency showed signs of compromise.
- While they were **orienting** and incorporating that second compromise, it escalated and cascaded further into the commonwealth infrastructure.

Initial signs showed there was a single compromised agency website. Other agency websites started showing similar signs of compromise. Analysts were unable to determine the source of the compromise, but indications showed it was cascading to other sites as well. Using all the available information a **decision** was reached to shut down (**act**) an entire state's web presence.

OODA was used until the root cause was discovered and the extent of impact was revealed. In each loop small decisions were made, action was taken, then observations began again to orient to the impact of those changes.

Root cause was eventually discovered to be a database compromised via an SQL Injection. A stored procedure that was used to serve up links was modified to present links to malware. Use of OODA allowed for a measured response while quickly determining the best path to resolution and restoration of web services.

That forced decision because if no action was taken, everyone coming to the commonwealth website would be served malicious links. The decision was to shut down entire state web presence. You may have to go through this loop repeatedly.

## 10.7    CASE STUDY 6: THE COLONIAL PIPELINE: THREE ASSUMPTIONS YOU SHOULD NEVER MAKE ABOUT RANSOMWARE[17]

### 10.7.1    BACKGROUND

On May 7, 2021, the Colonial Pipeline company proactively shut down the pipeline system in response to a ransomware attack (https://www.energy.gov/ceser/colonial-pipeline-cyber-incident). It wasn't until May 13, 2021, that the systems were restarted and delivery resumed. During that time at least eight states experienced significant gas shortages in the Southeast (https://www.cspdailynews.com/fuels/colonial-pipeline-cyberattack-causes-gas-shortages).

Colonial engaged cybersecurity firm Mandiant to assist with the incident response. Mandiant reported that the hack occurred when the bad actors used a credential (user name and password) to log in to a remote access system, which then gave them access to Colonial's business systems. Fortunately, the operational technology (i.e., systems that run the pipeline) was never accessed. (https://www.bloomberg.com/news/articles/2021-06-04/hackers-breached-colonial-pipeline-using-compromised-password)

Colonial Pipeline CEO, Joseph Blount Jr., testified to Congress that he chose to pay the $4.4M ransom 1 day after the attack (https://www.cnbc.com/2021/06/08/colonial-pipeline-ceo-testifies-on-first-hours-of-ransomware-attack.html). He also testified that the pipeline was shut down voluntarily by "the imperative to isolate and contain the attack to help ensure the malware did not spread to the operational Technology Network."

At the time of the attack, Colonial did not have a Chief Information Security Officer or even a Cyber Security Manager. They were in the middle of recruiting a manager when the attack occurred. (https://www.cnn.com/2021/05/12/tech/colonial-pipeline-cyber-security-manager-job-search/index.html)

Could the $4.4 million ransom (https://www.cnn.com/2021/05/19/politics/colonial-pipeline-ransom/index.html) could have been avoided? Were signals that could have detected bad actors were targeting the U.S. oil giant? The short answer: Yes. But as debilitating as this attack was, it isn't the first "wake-up call" sounding alarms across the nation, nor will it be the last. As a country that's witnessed nearly 1,000 **ransomware attacks** (https://www.newyorker.com/news/daily-comment/the-colonial-pipeline-ransomware-attack-and-the-perils-of-privately-owned-infrastructure), on our critical infrastructure in less than a decade, we can't seem to get ahead when playing cybercriminals' game.

Many of you are probably reading this and thinking, "I'm not worthwhile to bad actors, they'd never spend their time trying to infiltrate my network," or maybe even, "My **cyber rating** says I'm in good health and therefore I have nothing to worry about."

Here are three of the worst assumptions you can make as a risk professional:

**#1: ASSUMING BAD ACTORS HAVE A RHYME OR REASON**

It's no wonder as to why a **cyber-attack** against a U.S. oil company would stir up some geopolitical buzz. In full transparency, although we do not have the confidential information necessary to confirm whether the attack on Colonial Pipeline was influenced by a foreign entity, but if you've been around long enough you would believe that's not the case. Instead, what *does* sit on top of a bad actor's checklist is ease of access. Colonial does not have a CISO, and as recently as a few months ago, Colonial Pipeline was seeking a cyber security manager. It doesn't take being a bad actor to understand why, without a CISO overseeing operations, a company of that size and influence would become a prime target for **ransomware**.

To make matters worse, ransomware now has an ecosystem of its own. That means that an amateur bad actor could have "tested the waters" by deploying a script to test whether they could access an open port. Figure 10.1, "digital asset discovery," shows the vulnerable asset of Colonial Pipeline that was most likely used in the attack. Once connected, elementary-level criminals can then employ the help of **Ransomware as a Service** (RaaS), to do the real damage.

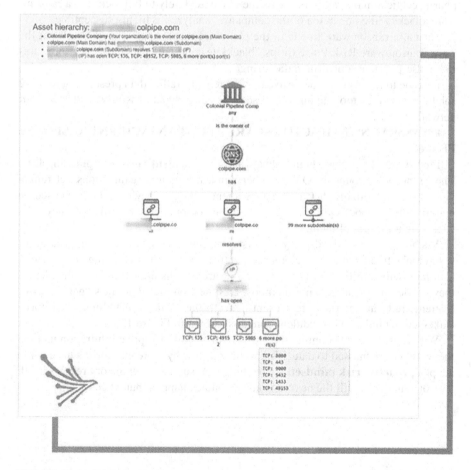

**FIGURE 10.1**   Digital asset discovery.

### #2: ASSUMING BAD ACTORS HAVE A MORAL COMPASS

Ransomware groups all have one thing in common: They're financially moti-vated. Where that money comes from, however, doesn't necessarily influence the decision on who to target. Case-in-point being recent attacks against the Irish health system (https://apnews.com/article/europe-asia-health-technology-business-2cfbc82beb75dfede32fc225113131b3), Scripps and, Care New England (https://healthitsecurity.com/news/care-new-england-resolves-weeklong-cyberattack-impacting-servers).

Healthcare has become a prime target for ransomware attacks because of what's at stake if the **cybercriminals'** demands are left unfulfilled. Can you imagine, how-ever, what would happen if, instead of cutting off an oil supply, pharmaceutical man-ufacturers were unable to distribute necessary drugs, or hospitals couldn't administer life-saving treatments?

Unfortunately, these scenarios aren't all that farfetched. In fact, Black Kite's **Ransomware Susceptibility Index**™ revealed that more than 12% of the top 200 pharmaceutical manufacturers across the globe are likely to fall victim to a ransom-ware attack, while one in ten of the companies analyzed is highly susceptible.

For more ransomware trends in the pharmaceutical supply chain, check out the 2021 Ransomware Risk Pulse (https://blackkite.com/whitepaper/2021-ransomware-risk-pulse-pharmaceutical-manufacturing/).

It's time to focus on our defensive risk gameplans, rather than prescribing reactive solutions when it's too late and a bad actor has already successfully infiltrated your network.

### #3: ASSUMING BAD ACTORS CARE ABOUT AN INCIDENT RESPONSE PLAN

First things first: You should always have an **incident response plan** in place. There's no way around it. On the other hand, too many organizations get tunnel vision when organizing their risk management strategy. They're so laser-focused on maintaining a "good" cyber rating, or meeting compliance standards that they miss what's right in front of them.

Visibility of your third-party ecosystem is important but knowing what your world looks like from a bad actor's perspective is crucial to defending your system. Common vulnerabilities present in ransomware attacks aren't a mere coincidence, they're tried-and-true hacking methods. Despite Colonial Pipeline's "good" **cyber hygiene**, red flags alerts were present beneath the surface, including: open ports, leaked credentials, and fraudulent domains (shown in Figure 10.2).

With ransomware becoming the most frequent and disruptive infiltration method the world has witnessed to date, it's a wonder as to why more enterprises have yet to adopt a **proactive risk mindset.** If one thing's for sure, it's that history repeats itself. It's not a matter of will the next ransomware attack happen, but when.

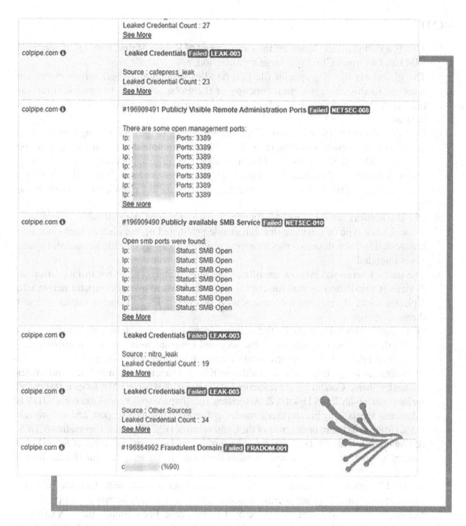

**FIGURE 10.2**   Colonial pipeline open ports, leaked credentials, and fraudulent domains.

## REFERENCES

Burson, S. January 19, 2010. Outsourcing information security. *CIO Magazine.*

Condon, R. December 5, 2007. How to mitigate the security risks of outsourcing. *ComputerWeekly.com.*

Kerth, N. n.d. An approach to postmorta, postparta, and post project reviews. http://c2.com/doc/ppm.pdf (retrieved February 12, 2013).

Ponemon Institute, LLC. March 2012. 2011 Cost of data breach study.

Schwartz, M. June 29, 2012. LinkedIn breach: Leading CISOs share 9 protection tips. *InformationWeekSecurity.*

*US v. Duronio.* Indictment USAO#2002R00528/JWD United States District Court, District of New Jersey.

## NOTES

1  This is a fictional case based on the experiences of Robert Maley, founder of Strategic CISO and a former CISO of a large organization.

2  The global.asa file is a special file that handles session and applications events on a server. In this case, the main function of the globa.asa file is to provide information to the web page regarding where the information that needs to be displayed is located.

3  By "remotely controlling the server," David is referring to the server being part of a botnet. This was discussed in Chapter 1. It is important to note that not all botnets can be detected with antivirus software. One may need a specific application that specifically removes that type of malicious software. Therefore, just because David did not find it with antivirus software does not mean that there was not any malicious software on the system.

4  P2P is a network where users can connect to each other and share files.

5  A rootkit is a type of software that can enable privileged access such as back door into a system. The back door is a way to access a system in a way that the site administrator never intended.

6  The terms *worms* and *viruses* are often used interchangeably but are in fact different. A virus is distributed by making copies of itself. A worm uses a computer network to replicate itself. It searches for servers with security holes and makes copies of itself there.

7  "Forensically sound" refers to the manner in which the electronic information was acquired. The process ensures that the acquired information is as it was originally discovered and thus reliable enough to be evidence in a court proceeding.

8  The information from this case was provided by Keith J. Jones: the court indictment and articles written by Sharon Gaudin (all are referenced at the end of the chapter). Mr. Jones is owner and senior partner with Jones Dykstra & Associates, Inc. (http://www.jonesdykstra.com/). JDA is a company specializing in computer forensics, e-discovery, litigation support, and training services. He is on the board of directors of the Consortium of Digital Forensics Specialists (CDFS; developing standards for the digital forensics profession). He is also the author of *Real Digital Forensics: Computer Security and Incident Response* (2005) and *The Anti-Hacker Toolkit* (2002).

9  A "PUT" option is purchased when someone thinks a stock will decrease in value by a certain date. In other words, it is essentially a contract between two parties to exchange an asset at a specified price by a certain date. For example, party A can purchase the stock at the decreased rate (specified in the contract) and sell it at the strike price (specified in the contract). The profit = (strike price)—(decreased rate)—(the cost of the PUT option). If the stock does not decrease in value, party A loses the cost of the PUT option.

10  The loss included $898,780 on servers, $260,473 on investigative services, and $1,987,036 on technical consultants to help with the recovery.

11  Root user is a special user account on a UNIX system with the highest privilege level.

12  A back door refers to an unauthorized way to access a computer system.

13  Swap space is where inactive memory pages are held to free up physical memory for more active processes.

14  Redirect questioning is the part of the trial process where the witness has an opportunity to refute information that may have damaged his or her testimony.

15  Red herrings are issues that are distractions to the real issue.

16 After 3 years of analyzing the UBS data, forensics expert Keith Jones came up with five points that helped UBS recover, as well as five points that will help them in the future.

17 Parts of this case are excerpts from Maley, B., "Three assumptions you should never make about ransomware," May 20, 2021, https://blackkite.com/three-assumptions-you-should-never-make-about-ransomware/

# Index

Note: **Bold** page numbers refer to tables; *italic* page numbers refer to figures and page numbers followed by "n" denote endnotes.

Printed in the United States
by Baker & Taylor Publisher Services